2021 AIChE Annual Student Conference Virtual Events

Held at the 2021 AIChE Annual Meeting

Boston, Massachusetts, USA and Online
7 - 11 November and 15 - 19 November 2021

ISBN: 978-1-7138-5722-8

Printed from e-media with permission by:

Curran Associates, Inc.
57 Morehouse Lane
Red Hook, NY 12571

Some format issues inherent in the e-media version may also appear in this print version.

Printed with permission by Curran Associates, Inc. (2022)

For permission requests, please contact AIChE
at the address below.

AIChE
120 Wall Street, FL 23
New York, NY 10005-4020

Phone: (800) 242-4363
Fax: (203) 775-5177

www.aiche.org

Additional copies of this publication are available from:

Curran Associates, Inc.
57 Morehouse Lane
Red Hook, NY 12571 USA
Phone: 845-758-0400
Fax: 845-758-2633
Email: curran@proceedings.com
Web: www.proceedings.com

TABLE OF CONTENTS

637107 Chemical Engineers in the Community

Abstract:

Engaging your community is a great way to encourage middle school and high school students to get excited about chemical engineering, STEM, and your institution! Stop by to see how the University of Michigan is creating connections with the local middle and high schools.

637104 Fostering a Chemical Engineering Community

Abstract:

Having a close knit ChE department is imperative to student success. Come check out how the University of Michigan fosters a community between students, professors and faculty. Topics will include, but are not limited to, professor/faculty luncheons, research lab tours, and integrated undergrad/grad events.

637140 Diversifying Technical Projects: The Chemical Engineering Introductory Project

Abstract:

The Chemical Engineering Introductory Project (ChIP) is a technical project that allows participants to learn about fundamental chemical engineering principles and apply them by designing small-scale continuous process systems in a team-oriented environment. In past years, our members have built coffee machines, self-sustaining boats, cooling towers, and soap batch reactors as their final project. ChIP is aimed toward incoming freshmen and transfer students, and since its introduction four years ago, AIChE at UCLA has seen an increase in student participation in technical projects and other areas of the club. The main challenge over the past year was adapting a traditionally in-person project to an online-only format while maintaining exposure to technical skills, a team oriented environment, and participation among new students. Topics covered in this workshop will include designing new projects, connecting small-scale projects to core engineering fundamentals, and building community amongst chemical engineering students.

637141 Lifetime Engineering Applications Project: LEAPing into a lifetime of learning

Abstract:

LEAP, the Lifetime Engineering Applications Project, is a technical project started by students who are part of AIChE at UCLA. Students learn how to use process design simulation software, called AVEVA PRO/II, to model real industrial processes while also taking economic analyses and ethics into consideration. We have developed a variety of unique projects to expose students to many aspects of chemical engineering, such as alcohol distillation, ammonia production, crude oil processing, biofuel synthesis, and perfume manufacturing. Although this project has a low cost, students complete it with a deeper understanding of chemical engineering fundamentals and preparation for senior design projects as well as industry challenges. The project leads sharpen their presentation skills and gain experience in preparing lesson plans while having fun. In this workshop, current LEAP leads will discuss their experiences and how this project has also been continued virtually.

637109 Making the Most of Your Network: How to Form Connections with Alumni and Start an Alumni Mentorship Program

Abstract:

As college students, we are often told that our years in school are essential to building our network. But how can you form these valuable connections and how far could your network extend? This workshop will go beyond connecting with your classmates and professors and delve into the world of alumni networking. Many schools maintain sums of alumni who are waiting to help their alma mater, but they are often an untapped resource for students. Northeastern University AIChE has created a unique alumni mentorship program that pairs students with an alumni mentor based on similar interests and career goals. We will teach you how to establish meaningful connections with alumni, solicit a potential alumni mentor, and start an AIChE chapter alumni mentorship program. We will also go further into how to maintain these connections and discuss all the benefits your university alumni have to offer. Please join us for this essential, interactive workshop in order to make the most of your network.

637112 Leveraging Existing Resources to Increase Involvement and Benefit Membership

Abstract:

Officers from NC State's student chapter of AIChE will describe their initiatives to increase involvement and benefit membership. The initiatives include implementing events for casual conversations between undergraduates and professors, leveraging the alumni network for mentorship opportunities and speaking engagements, and utilizing the Engineering Career Fair to gain new sponsorships. They will describe how to identify and take advantage of existing frameworks available at your university.

637111 Freshmen and Transfer Involvement: AIChE at UCLA's Inaugural Intern Board

Abstract:

Incoming freshmen and transfer chemical engineering students are often thrust into the complex and unfamiliar world of engineering without a clear idea of what to expect or how to make connections (not to mention in the unprecedented state of remote learning). Officer service is a great way for students to gain invaluable leadership and mentorship experience, building up their professional accomplishments and profile. Here, we plan to discuss the UCLA AIChE Student Chapter's newly conceived Intern Board (founded 2020), where freshmen and transfer chemical engineering students supported more senior members over a six-month assignment. To improve how we engage with students entering UCLA who are interested in becoming more involved with our chapter, we implemented the Intern Board to be N.I.C.E.

- Networking-friendly by allowing direct mentorship between department heads and their interns as a smaller-scale unit than in committees
- Inclusive of club departments that have no history of committees
- Cohesive in new member identity-- with a shared intern identity and social intern events
- Expansive in our event offerings and professional initiatives by allowing interns to directly partake in their development

We will first begin with a description of the pre-existing committee model, and cover the associated issues of engagement and retention. We will then proceed to discuss the areas interns served in: Internal Events, Treasury, Alumni & Outreach, and Publicity.

We will also be going over intern testimonials, the continuity of the interns' involvement with AIChE after the end of their internship, and future directions we will pursue based on feedback.

637106 Remote Reach: Innovating K-12 Education in a Time of Social Distancing and Beyond

Abstract:

In the face of the COVID-19 crisis, parents and educators were tasked with the enormous responsibility of creating a dedicated learning environment for their students at home. While shifting to an online medium was challenging, it provided a unique opportunity for AIChE at UCLA to support K-12 students through its new Remote Reach project. In Remote Reach, designing and delivering personalized interactive lessons to students has proved to be a rewarding and enjoyable experience for members involved. The prospect of online education has enabled our chapter to establish several connections with schools and community groups and develop a new approach to outreach. This year, our chapter's Remote Reach, was a year-long project that was dedicated to helping elementary and high school students learn and get excited about STEM topics while learning from home. We partnered with 3 main schools, 2 elementary schools and 1 high school. For each elementary school, we had a team of 5 - 10 AIChE members make and present a unique STEM module biweekly to 100+ students. Topics were chosen based on teacher, school, and member input and were designed to include demos, videos, and interactive content. The goal was to facilitate learning and discussion amongst the students using different mediums, rather than just our members lecturing them. These biweekly modules gave the elementary students the opportunity to learn new topics on a regular basis and become comfortable with our members. For the high school, we designed modules that would teach them about what ChemE's do and how wide-ranging our careers can be with videos and an interactive Process Flow Diagram activity. At this workshop, we will introduce innovative ways our chapter has served hundreds of students in the Greater Los Angeles community online, and how all student chapters can grow their own outreach programs in this ever-changing climate. Additionally, we will detail how Remote Reach's techniques can be utilized in both an in-person and online format to best serve your community as guidelines change.

637103 The Importance of Philanthropy

Abstract:

The chapter at Purdue University is known for its unique relationship with the surrounding community. Our philanthropy committee holds events that are highly sought after by our members and provide impactful services to organization in the Lafayette community. We have a unique relationship with one of our industry sponsors where we will collaborate on a philanthropy event and build houses in the community. The Outreach K-12 Program hosts events at local schools and educates elementary, middle school or high school students on Chemical Engineering Fundamentals. What I would like to discuss is how the Purdue Chapter's philanthropic and outreach events engages our members and educates them on the importance of giving back to a community and inspiring a younger generation. Next after showing how Purdue plans philanthropy events, I would like to open up a discussion on how the audience can get involved in their communities or plan outreach events. I believe that philanthropy and outreach events enrich members with communication skills, relationship building, and the ability to motivate others.

637113 Developing Your Mentorship Program

Abstract:

Mentorship is a great way to involve and develop underclassmen in your chapter while fostering a tight-knit community within your ChE department. Come learn how the University of Michigan AIChE Chapter runs their undergraduate and graduate mentorship programs!

638250 Moving at the speed of science: The development of the Pfizer/BioNtech mRNA vaccine for the Covid-19 pandemic

Abstract:

In March of 2020, as Covid-19 infection was spreading around the world, Pfizer joined forces with BioNtech to co-develop BNT162, a messenger RNA (mRNA) vaccine candidate for the potential prevention of Covid-19 infection. This was the second such collaboration between the two companies with the initial one focused on using the same mRNA technology platform to develop flu vaccine. However, in the case of the Covid vaccine development, the urgency was utmost as lives were being lost around the world due to the virus and there was no viable treatment available. The Covid pandemic was arguably the biggest challenge facing humanity in a generation.

Pfizer and BioNtech agreed to leverage the full resources and expertise of both companies to accelerate the development of the vaccine. The project was termed lightspeed with a highly ambitious goal of developing and gaining authorization/approval within a year. This seemed like an impossible task especially since the average time for vaccine development is typically 10-15 years and the fastest any vaccine had ever been developed was 4 years (mumps vaccine). However, within about 9 months of development, the safety and efficacy of the Pfizer-BioNTech Covid-19 vaccine had been established in a pivotal phase 3 study and emergency use of the vaccine authorized by the FDA. The talk will describe how courageous leadership, strong collaboration, and a new mRNA technology platform enabled the accelerated development of the Pfizer-BioNTech Covid-19 vaccine.

636762 Checklist for Life After Graduation

Abstract:

So you've been a student for the last 16 years of your life. Crazy, right? Graduating from college is a very important, exciting, and sometimes daunting transition period. In this session, you will hear from two recent ChemE grads who will share their tips on how to make the transition to adult work life as seamless as possible.

637687 Finding the Right Company through the Questions You Ask

Abstract:

Interviewing for a full-time job opportunity after graduation can be a challenging experience for a number of reasons. Most often it is challenging because companies invest a good bit of time into finding the right candidates through specific interview processes. But it is just as important for you as an interviewee to invest a good bit of time into finding the right opportunity for you. One way to do this is through the questions you ask the interviewers. Come listen to some highlights on Evonik's interview process as well as some tips on how to find the right company for you.

636763 Getting your PE License

Abstract:

This presentation discusses how becoming a Professional Engineer (PE) can enhance a chemical engineer's career. Students learn about the PE licensure process and the two exams required to become licensed. The Computer Based Testing (CBT) exam content for both the Fundamentals of Engineering (FE) Exam and Practice of Engineering (PE) exams are described. The presentation utilizes embedded videos of young PEs and stresses the value of taking the FE exam before, or shortly after, graduation. It aims to demystify the FE/PE process. The workshop leaders are Joe Cramer and Bill Parrish who are chemical engineering PE's and active members of the AIChE's Licensing and Professional Development Committee as well as the National Council of Examiners for Engineering and Surveying's Chemical PE Exam Committee. Both have also served on AIChE's Career and Educational Operating Council which oversees licensing for AIChE. The workshop encourages Q&A and interaction with attendees.

636767 Custom-Built Thin Films – A Molecular Dynamics Perspective

Abstract:

The workshop will introduce participants to multi-scale modeling, with the focus on molecular dynamics as a tool for custom-building materials for several applications including electronics, drug delivery and industrial processes. Steps necessary to develop representations of physical systems will be outlined and implemented within the LAMMPS software application for a simple example during the workshop. Students will learn how to analyze output and visualize results for the above demo. Using the demo as a framework, a more complex, industrially relevant example will be discussed to further demonstrate the breadth of capabilities of the molecular dynamics technique. The workshop will wrap up with a brief discussion on the recommended type of graduate training for a career in engineering custom-built materials using multi-scale modeling. The syllabus for the workshop is provided below:

- High-level discussion on the use of multi-scale modeling of materials
- Different methods presented in terms of their time and length scale
- Molecular dynamics background and description
- Building models – from the actual system to a model of it
- Definition of interactions – what can you simplify, what do you need to account for
- Getting it done – LAMMPS input
- Getting your results
- Visualizing results
- Demo: a small 5 minute problem (what we learn?) – live action
- Results from a problem relevant to custom-building of thin films
- Resources required to support this work: high-performance computing
- Requisite skills and graduate training

636765 Non-Traditional ChemE Careers

Abstract:

Why did you become a ChemE? Many students say they liked chemistry in high school, were good at math or science, or saw that ChemE was one of the highest paid engineering disciplines. But what CAN you do with your degree? In this session, you will hear from a panel of ChemEs that chose an unconventional route after graduation and what led them down the path that they chose.

636766 The Importance of Teamwork and Collaboration

Abstract:

We all work in teams these days, whether we work in an industrial, academic, government or non-profit organization. This session will focus on a guided discussion on the importance of teamwork and collaboration to achieve and maximize results. You will hear from managers about the importance of creating and working in an environment that encourages everyone to work together as it relates to team's results. Finding the correct balance between autonomous work, teamwork and collaboration skills is critical for workforce engagement and for team's performance.

636761 Mixing of Viscous, Non-Newtonian Fluids

Abstract:

Many viscous fluids processed in industry exhibit non-Newtonian rheology. These industries include paints, personal products such as shampoo and detergents and food products such as ketchup and mayonnaise. All at some stage in their production require mixing. Most of these fluids are shear-thinning which means that the apparent viscosity of the fluid will be low near the impeller, where velocity gradients are high, and high near the vessel wall where velocity gradients are low. This presentation will discuss how to select the impeller type that is appropriate for a particular operating regime (Reynolds number). Then correlations for estimating blend times in Newtonian fluids will be discussed and how they can be modified to take account of non-Newtonian fluid behavior. There is a class of fluids which exhibit a yield stress, meaning that a minimum level of shear must be applied before the fluid will move. Examples include toothpaste and high solids loading slurries. The rules used to size mixing equipment for these fluids will be discussed.

636769 Navigating Dynamic Careers in Catalysis and Reaction Engineering

Abstract:

Panelists will discuss their careers in the field of Catalysis & Reaction Engineering; how the Catalysis & Reaction Engineering Division has played a role in their careers; and advice on how to navigate a dynamic and fulfilling career in the field. The panel will be designed to engage the audience and provide opportunities for interaction.

636813 Applying to Graduate School and Research Fellowships

Abstract:

Why should you pursue a graduate degree in engineering? What should you consider when applying to graduate school? How do you ace your fellowship and graduate program applications? How can AIChE help you prepare for graduate school and excel as a graduate student as you pursue your degree? Members of AIChE's Young Professionals Committee (YPC) will be discussing answers to all these questions and more. Join us for a great session discussing everything you need to prepare for your graduate school career.

This workshop will feature numerous YP's either currently in graduate programs or recently graduated. This session has been very successful in the past, as numerous students attending the AIChE conference are interested in pursuing research-based graduate degrees in engineering. The session will be interactive, leaving ample time for Q&A. Furthermore, this session can be easily adapted to a virtual format if need be.

One of the society-wide goals of AIChE is to add new members and increase member retention. This session will also focus on how AIChE can help students prepare for graduate school and succeed once they start to pursue their degree by advertising for graduate-student leadership in YPC and potentially divisions and forums.

636760 Exploring the Wild World of Nanoscale Science and Engineering: From School to Industry

Abstract:

The world of nanoscale materials has become vast and overly complex. Numerous avenues for research, development, and application have made understanding the impacts of the world of nano more complicated. Whether it is nanobiotechnology or energy-related fields, nanoscale materials have become an important thrust for future growth of science and technology. In this workshop, we will delve deep into the world of nanoscale science and engineering. Examples from nanobiotechnology, carbon nanomaterials, and nanomaterials for energy-related applications will showcase how you can be a part of this ever-growing world. An interactive discussion panel will help audience members understand the different career paths available for pursuing work involving nanoscale materials. It is expected participants will leave with a broader and fuller understanding of nanoscale science and engineering and the importance of their contributions to its continual evolution.

637682 How to Make the Most of your Omega Chi Epsilon Chapter

Abstract:

Officers and members of Omega Chi Epsilon, the National Chemical Engineering Honor Society, are encouraged to attend this session to meet with national officers and share best practices and favorite activities. Students from schools that do not have an Omega Chi Epsilon chapter are also invited to attend to learn more about establishing a chapter on their campus.

638673 TEM Characterization of the Structure and Composition of Nickel-Iron Catalysts for Alkaline Oxygen Evolution Reaction

Abstract:

Alkaline anion exchange membrane (AAEM) electrolyzers have high potential to support renewable energy systems by storing energy in the form of hydrogen. However, the sluggish kinetics of the anodic half reaction, the oxygen evolution reaction (OER), significantly hinders the efficiency of these devices. Large overpotentials of 300 mV or more are needed to overcome the slow kinetics of the OER, which diminishes the feasible efficiency of this process by tens of percent. Ultimately, these higher energy barriers translate to financial costs and limit the practicality of AAEM water electrolysis.

Effective electrocatalysts can lower the overpotential for OER, but these catalysts need to be highly active, stable for thousands of hours, and inexpensive. The state-of-the-art catalysts for the OER in the current generation of AAEM electrolyzers are IrO_x and RuO_x; however, these catalysts contain precious transition metals that are prohibitively costly to deploy on a large scale. Oxidized nickel-iron based catalysts ($NiFeO_x$) are promising non-precious alternatives, which have shown even higher activity than IrO_x in alkaline conditions. However, the long-term stability of these materials has not been adequately demonstrated to date.

We have synthesized carbon-supported NiFe alloy using a facile wet-impregnation method, which is electrochemically pretreated to generate a catalytically active $NiFeO_x$ phase. We have shown that these alloy-derived NiFeOx catalysts achieve a current density of 10 mA/cm^2 at approximately 280 mV of overpotential at room temperature and at 240 mV of overpotential at 70°C in alkaline aqueous solution. However, the stability of this catalyst in concentrated alkaline electrolytes is ambiguous. Moreover, there is evidence that after spending up to several hundred hours in a colloidal suspension with Nafion ionomer and isopropyl alcohol the catalyst undergoes structural and compositional changes that lead to unpredictable catalyst stability. Ongoing work is focused on utilizing identical location transmission electron microscopy techniques to investigate how the structure and composition of the catalyst changes throughout its lifetime and impacts catalyst activity and stability.

Title: Catalytic Methane Dehydroaromatization: Determination of the Structure of Pre-Catalysts and Effect of the Synthesis Protocol on Structure & Activity of Promoted Catalysts

Natural gas, primarily composed of methane, is an abundant resource. Unfortunately, due to its abundance and the lack of economical transport infrastructure, billions of cubic feet of natural gas are flared yearly, wasting a valuable resource and releasing CO_2 in the process[i].

The methane dehydroaromatization (MDA) reaction provides a potential solution. MDA converts methane to aromatic compounds such as benzene in one step, thus enabling on-site processing and providing an excellent opportunity of distributed manufacturing of aromatics while minimizing transportation costs.

Mo/ZSM5 catalysts have shown promise as MDA catalysts, but further investigation is needed to understand the mechanisms of this catalyst and improve its performance. Two areas not yet fully understood are explored here: (1) the structure of the Mo oxide species in the pre-catalysts, (2) the effect of adding an iron promoter to the Mo/ZSM-5 catalysts.

While some research suggests that a $(Mo_2O_5)^{2+}$ dimer exists[ii], other studies show that the MoO_4 monomer is the sole MoO_x species[iii]. Here we perform temperature-programmed oxidation (TPO) of the catalyst precursors in order to determine the stoichiometry of the MoO_x species, which will help distinguish whether monomer and dimer species are predominant in the pre-catalysts

Additionally, the physicochemical properties of Mo-Fe/ZSM-5 catalysts are examined. Prior studies suggest that impregnation of a secondary metal can improve the performance of Mo/ZSM5 catalysts[iv]. Here we explore various synthesis protocols to add Fe promoter to the Mo/ZSM-5 catalysts and we study the effect of the protocol on the catalyst structure and activity.

---

[i] US EIA, Natural Gas Annual, 2018.

[ii] R.W. Borry, Y.H. Kim, A. Huffsmith, J.A. Reimer, E. Iglesia, J. Phys. Chem. B 103 (1999) 5787–5796.

[iii] J. Gao, Y. Zheng, J.-M. Jehng, Y. Tang, I.E. Wachs, S.G. Podkolzin, Science (80-.) 348 (2015), 686 LP – 690.

[iv] A. Sridhar, M. Rahman, S.J. Khatib, ChemCatChem 10 (2018) 2571–2583.

638397 Role of Nitrogen Defects on the Activity of Fe-N-C Catalysts Towards the Oxygen Reduction Reaction

Abstract:

Hydrogen fuel cells are electricity-generating devices that produce clean energy with no SO_x or NO_x emissions. The oxygen reduction reaction (ORR) occurs at the cathode in hydrogen fuel cells and produces water as the major product. Currently platinum and platinum alloys are the best electrocatalyst for ORR but are expensive. Fe-N-C catalysts are a class of non-precious group materials that are promising for ORR as they show reasonable activity compared to platinum. They are synthesized by pyrolysis at about 850°C, which leads to production of several different moieties in the catalyst including iron nanoparticles in the form of nitrides, and carbides, FeN_x (x=1-4) moieties, and nitrogen moieties. A range of nitrogen defects have been identified experimentally, including pyrrolic, pyridinic, graphitic, protonated, and cyanide nitrogens with total concentrations typically ranging between 5-30 atomic %. While FeN_x sites are thought to show highest contribution to ORR activity, nitrogen defects have also been proposed to play a role in ORR. FeN_x sites are proposed to reduce O_2 to H_2O, graphitic nitrogens and protonated pyridinic nitrogens are proposed to reduce O_2 to H_2O_2, while pyridinic nitrogens are proposed to reduce H_2O_2 to H_2O. The aim of this study is to understand the role of these nitrogen moieties in ORR.

To begin, graphitic nitrogen and protonated pyridinic nitrogens were considered in varying concentrations and locations in a graphene framework. Then, they were considered in a graphene framework along with FeN_4 sites in the bulk of its sheet and its edge. First principles density functional theory (DFT) was used to model these defects and find energetics for nitrogen incorporation and that for ORR. Preliminary results for the graphene case show that protonated pyridinic nitrogens at the edges have lower nitrogen incorporation energies compared to graphitic nitrogens in the bulk, making edge sites more preferrable. These nitrogen defects may also influence where ORR intermediates bind on the surface due to change in the electronic structure of the system. Currently, we are probing this aspect and extending our analysis to the cases where FeN_4 sites are also present in the framework. This analysis will contribute towards improving the understanding of the complex structure of the Fe-N-C catalyst, which is important to improve ORR activity and catalyst stability for fuel cell applications.

638476 Plasmon Mediated Synthesis of Titanium Nitride-Nickel Nanocomposite and Their Application in Photocatalytic Reduction of Bicarbonate

Abstract:

Plasmon enhanced catalysis is a sustainable, energy-efficient means of triggering a reaction through utilization of solar light. Refractory stable plasmonic nanoparticles such as titanium nitride (TiN) can effectively absorb solar light and convert that into heat or generate photoexcited charge carriers. A composite catalyst developed combining the plasmonic properties of TiN with transition metal nanocatalysts Ni or Pt can effectively drive a wide range of catalytic reactions using solar light. To investigate this effect, TiN-Pt and TiN-Ni nanoparticles were synthesized through photodeposition of Ni and Pt onto TiN. The composition of the composite nanoparticles were confirmed using UV Vis spectroscopy, HRTEM, EDX, XPS and XRD. The catalytic effect of these composite catalysts was determined by their ability to aid in the reduction of bicarbonate to formate with glycerol as the hole-scavenger. It was found that TiN-Ni enhanced that particular reaction better than TiN-Pt; although, TiN-Pt also achieved greater conversion than undoped TiN. The findings of this study can be applied to develop other plasmonic catalysts which can successfully enhance different energy extensive reactions. The work is funded by ACS PRF (# 60878 UR-6).

638336 High-Valent Polyoxovanadate-Alkoxide Clusters Synthesis and Their Use in the Dehydrogenation of Alcohols

Abstract:

Polyoxovanadates (POVs) have drawn attention due to their potential in many fields including biological, energy, and catalysis innovations. This project investigates the potential use of POV clusters in the dehydrogenation of alcohols under UV light. The methyl-backed variant ($[^{n}Bu_4N]_2[V_6O_{13}(TRIOL^{Me})_2]$) and the nitro-backed variant ($[^{n}Bu_4N]_2[V_6O_{13}(TRIOL^{NO2})_2]$) are synthesized from commercially available starting materials like Sodium Orthovanadate (Na_3VO_4), tetrabutylammonium bromide, and the appropriate TRIOL ligand precursor. The alcohols were used to test the clusters' strength and ability to form the corresponding ketones and aldehydes. Various alcohols, including phenylethanol, isopropanol, pentanol, methanol, and ethanol, have been chosen based on previous research done in this field. Reaction products were characterized using ^{1}H NMR Spectroscopy and Electronic Absorption Spectroscopy. NMR peaks of the products were then compared to previously published data. Results showed the ability of the clusters to oxidize some alcohols under UV light after a few hours/days like the oxidation of 1-phenylehtanol to acetophenone whereas some alcohols, such as methanol, ethanol, showed slight/ no signs of getting oxidized. The project has room for improvement by choosing a wider range of alcohols and relating the alcohols' structures to their potential to be dehydrogenated. Future work will investigate the potential of creating a catalytic cycle by exposing the reduced clusters to O_2 and converting them to their original oxidized form.

638253 The Effects of Alkali Metal Cations on Oxygen Evolution in Acidic and Basic Electrolytes

Abstract:

The choice of electrolyte in electrocatalytic systems can significantly impact measured performance. While these effects are well documented, a clear explanation remains lacking. Here, we demonstrate that alkali metal cations have a promoting effect on the oxygen evolution reaction (OER) over iridium (Ir) catalysts, even when accounting for the confounding effects of pH. Additionally, we show that the cations' promoting effect is more prominent in relatively high pH levels and becomes insignificant as the electrolyte is made strongly acidic. In the absence of cations, we find a positive correlation between acidity and activity for OER over Ir catalysts. We hypothesize that cations favorably interact with reaction intermediates and help optimize interactions between the catalyst surface and key adsorbates. These experiments further our understanding of the role of the electrolyte in controlling electrocatalytic activity.

636941 Enhanced ORR Activity of Fepc Functionalized Graphene Via Substrate Doping and/or Ligand Exchange: A First Principles Study

Abstract:

Platinum, an expensive and highly precious metal, is currently used for both electrodes of a Proton-exchange membrane (PEM) fuel cell. The cathode reaction, in which the O_2 molecule is reduced (oxygen reduction reaction (ORR)), has been a major barrier in the development of economically enticing technology. High-performance catalysts fabricated with little to no platinum for the cathode of a PEM fuel cell remain a significant challenge for commercial application due to the sluggish kinetics. Because of this, there are many non-precious catalysts currently being investigated to find suitable alternative that offers better integrity and stability than the standard platinum-based catalysts. A single-metal atom fuel cell catalyst, iron phthalocyanine supported on graphene (GFePc) has showed promise as a potential alternative through its performance and stability. In the present work, ligand-exchanges (F-, NH_2-) of FePc and substrate doping (B-doping, N-doping) of graphene are explored to tune the performance of GFePc using spin-polarized density functional theory (DFT) calculations for the associative mechanism, where ORR and water formation reaction (WFR) occur. Additionally, increased doping percentages and a combination of a ligand-exchange and substrate doping were also explored. Some of the calculations performed in this study include ORR and water formation reaction (WFR) processes in mild practical fuel cell operating conditions, elementary electrochemical catalytic reaction steps, and free energy landscapes. Based on our calculations and analysis, ~1 at. % boron doping seems to improve the performance of GFePc by lowering the over potential required for ORR and WFR to take place to be ~0.63 eV.

Acknowledgements: The work is supported by ACS-PRF [58740-UR6] and used the Extreme Science and Engineering Discovery Environment (XSEDE) TACC at the stampede2 through allocation [TGDMR140131]. This work utilized resources from the University of Colorado Boulder Research Computing Group, which is supported by the National Science Foundation (awards ACI-1532235 and ACI-1532236), the University of Colorado Boulder, and Colorado. PCC Cluster, NM Consortium, NM.

637661 Synthesis and Characterization of Zeolite-Encapsulated Organometallic Complexes That Catalyze Selective Alkane Oxidation

Abstract:

Methane is an unreactive greenhouse gas that is abundant in shale gas. We hypothesize that metal phthalocyanines (MPCs) encapsulated in faujasite (FAU) zeolites (MPC@FAU) can selectively activate methane and other light alkanes, due to their similarities to metal-containing zeolites and MOFs that perform alkane oxidation with nitrous oxide (~393 K). Prior studies have observed heterogeneity in metal species present on existing catalysts, which renders it difficult to draw conclusions about desirable catalyst traits across varied metal identities without first developing careful active site quantification protocols. MPCs contain divalent central metal atoms bound to nitrogen atoms in square planar geometry. FAU zeolites are crystalline microporous aluminosilicates whose pore intersections form supercages (~1.2 nm diameter); these supercages host MPCs (~1.2 nm diameter), and this predictable structure establishes a controlled environment for experimental and computational catalyst characterization. However, undesired MPCs located on the exterior of FAU (MPC/FAU) are unavoidable during synthesis. We have developed a reproducible washing procedure to remove MPC/FAU species, enabling quantification of encapsulated active sites simply via bulk elemental analyses. Copper perfluorophthalocyanine ($CuF_{16}PC$), was added to FAU zeolites via both hydrothermal synthesis ($CuF_{16}PC$@FAU) and post-synthetic deposition ($CuF_{16}PC$/FAU). X-ray diffraction patterns and micropore volumes measured by nitrogen adsorption demonstrate that the obtained materials are crystalline FAU. Samples were washed by repeated vortex mixing and centrifugation using six different solvents. Using ultraviolet-visible spectroscopy (UV-VIS), the amount of $CuF_{16}PC$ in each supernatant was quantified, along with collection of electron paramagnetic resonance (EPR) and diffuse reflectance UV-VIS spectra of the recovered solids, to identify which solvent(s) removed the $CuF_{16}PC$/FAU species most effectively and how many washes were required. Ongoing studies of this tunable single atom catalyst platform will allow for determination of desired characteristics of alkane oxidation catalysts without the distractions of an ill-defined distribution of metal species.

637697 Modeling Ce^{3+} and Ce^{4+} Structures and Redox Kinetics in Acidic Electrolytes

Abstract:

The Ce^{3+}/Ce^{4+} redox couple is used in many fields, including organic synthesis, waste remediation, and energy storage. Despite the Ce^{3+}/Ce^{4+} redox couple's broad applicability, the structures of complexes in many acidic electrolytes remains unknown. Our recent work using UV-Vis spectroscopy and DFT calculations shows that Ce^{3+} likely forms a complex with nine waters, whereas Ce^{4+} favorably undergoes single ligand exchange in seven acidic electrolytes (HCl, H_2SO_4, H_3NSO_3, CH_3SO_3H, HNO_3, CF_3SO_3H, and $HClO_4$) [1]. Here we report new experimental EXAFS and molecular dynamics-based EXAFS of cerium complexes to further elucidate the structures of the cerium complexes in solution. In addition, the Ce^{3+}/Ce^{4+} redox couple provides a unique opportunity to study how changing ion complexation of a single oxidation state (Ce^{4+}) affects the electron transfer kinetics, while the other oxidation state (Ce^{3+}) remains a constant structure. We apply theoretical charge transfer models based on the structure of the cerium complex to understand what controls the Ce^{3+}/Ce^{4+} mechanism and compare calculated redox kinetics to experimental kinetic trends. We extract kinetic parameters of an outer sphere redox reaction using DFT calculations within a polarized continuum model framework. This research increases understanding of cerium ions in acidic electrolytes and tests theoretical paradigms for charge transfer.

1. Cailin A. Buchanan, Eunbyeol Ko, Spencer Cira, Mahalingam Balasubramanian, Bryan R. Goldsmith, and Nirala Singh *Inorganic Chemistry* **2020** *59* (17), 12552-12563 DOI: 10.1021/acs.inorgchem.0c01645

637767 In Situ Generation of Hydroxyl Radicals Using UV Light and TiO2 Photocatalysts on Quartz Felt

Abstract:

As water scarcity continues to become a growing concern, it is imperative that water purification technologies improve. Current best methods, such as advanced oxidative processes (AOPs), can be highly effective against targeted contaminants, but often require chemical additives (e.g. chlorine, peroxide, ozone) which increase costs and may lead to the formation of undesirable byproducts. Immobilized photocatalysts with strong light absorption, chemical stability, and high surface area could greatly reduce the need for oxidant feedstocks and improve the affordability of UV-AOPs. In this study, titania-based photocatalysts immobilized on quartz-fiber felt are used for UV-driven water treatment. Our mesoporous TiO_2 and macroporous quartz felt yield a high surface area beneficial for mass transport, and optimal optical path length for UV light (250-400nm). Au nanoparticles can be added to the titania to increase the generation of reactive oxygen species (ROS) and provide alternative reaction pathways for degrading contaminants. Long term stability (>24 hour) to degrade model contaminants was tested under UVA LEDs and germicidal UVC lamps. Our photocatalysts produce >100 μM *OH in 20 seconds in a flow reactor and achieve an Electrical Energy per Order (EEO) of <0.6 for the degradation of Rhodamine B. This technology has the potential to be scaled up and integrated into current water purification systems to increase efficiency.

638586 Carbon Fiber Paper Working Electrodes for CO_2 Reduction Electrocatalysis

Abstract:

Rapid climate change is threatening the world as we know it. Climate change is caused by global warming, primarily due to the release of greenhouse gases, such as CO_2. Carbon dioxide emissions continue to rise at an alarming rate by burning fossil fuels, cement production, and deforestation. If no further action is taken to curb continued anthropogenic CO_2 emissions, devastating impacts on the planet are inevitable, such as heatwaves, droughts, wildfires, and decline in water and food supplies. Successor technologies must be developed now to combat climate change; one such sustainable technology is aqueous electrocatalysis to convert carbon dioxide to valuable fuels and chemicals.

Inexpensive, electrically conductive, high surface area electrode materials, such as carbon fiber paper, are needed for globally scalable electrolyzers. A large obstacle to the wide-spread use of carbon fiber paper is its high hydrophobicity; electrocatalysis in aqueous media requires hydrophilic surfaces to maximize the contact area between nanocatalysts and the electrolyte.

Here we present methods to increase the hydrophilicity of carbon fiber paper. We systematically investigated how different treatments affected surface chemistries. We were able to show that distinct surface species correlated with the durability of hydrophilicity. We characterized carbon fiber paper samples by high-resolution X-ray photoelectron spectroscopy and scanning electron microscopy imaging with energy-dispersive X-ray spectroscopic elemental mapping.

We used treated carbon fiber paper to immobilize ca. 20 nm gold nanoparticles on these electrodes, using commercial aqueous citrate-capped gold colloid, for CO_2 reduction electrocatalysis. Scanning electron microscopy images showed that gold nanoparticles were evenly distributed on carbon fibers. Hydrogen was the predominant product in CO_2-saturated aqueous 0.1 M pH 6.8 $KHCO_3$ electrolyte; the only other detected product was CO. We note that citrate capping of gold nanoparticles enhances hydrogen production, which explains the observed high faradaic efficiencies for hydrogen. During chronoamperometry testing, currents initially decayed, likely attributable to the reduction of native oxide. After this induction period currents were stable, indicating that immobilization of gold nanoparticles on treated carbon fiber paper was successful.

638696 Electrochemical Characterization of Nife-Layered Double Hydroxide Electrocatalysts for Energy Storage Applications

Abstract:

Electrochemical Characterization of NiFe-Layered Double Hydroxide Electrocatalysts for Energy Storage Applications

Sean P. Rogers[1], Brady Weathers[1], Samuel Baldwin[1], Matthew DiBiase[1], Thomas Gascoigne[1], Jiangtian Li[2], Rongzhong Jiang[2], Deryn Chu[2*], Caspar C. Yi[1], Enoch A. Nagelli[1*]

[1]Department of Chemistry & Life Science, Chemical Engineering Program, United States Military Academy, West Point, New York 10996

[2]U.S. Army Combat Capabilities Development Command, Army Research Laboratory

2800 Powder Mill Rd., Adelphi, MD 20783-1107

Ni-based foam electrocatalysts are promising cost-effective alternatives to noble metal catalysts for oxygen evolution reaction (OER) in water-splitting. Ni-layered double hydroxides (LDH) and NiFe-LDH foams were synthesized with electrodeposition techniques with commercial Ni foam. Alternative catalysts such as 3D nickel foams with high surface area and surface-to-volume ratio are promising materials to reduce kinetic losses and enhance mass transfer rates. Raman Spectroscopy is used to confirm the presence defects from the layered double hydroxides with nickel. Scanning electron microscopy (SEM) and energy dispersive x-ray spectroscopy (EDX) were used to characterize the morphology of the 3D foam structure and the surface elemental composition. Electrochemical characterization techniques such as electrochemical impedance spectroscopy (EIS), cyclic voltammetry (CV), and potentiostatic measurements are used to characterize the areal specific resistance (ASR), areal capacitance, electrochemical surface area, and electrochemical stability, respectively.

KEYWORDS: Ni-foam catalysts, water-splitting, electrocatalysts, 3D Ni foams, Layered Double Hydroxide

CONTACT: Dr. Enoch A. Nagelli, Email: enoch.nagelli@westpoint.edu; Dr. Deryn Chu, Email: deryn.d.chu.civ@mail.mil ,

638417 Silica Particles As an Effective Scavenger for Platinum Catalyst Recovery

Abstract:

Hydrosilylation is the addition reaction between a silicon hydride and an unsaturated bond to form a silicon-oxygen or silicon-carbon bond. Platinum is usually used as a catalyst in hydrosilylation reactions for synthesis of organosilicon products i.e., organosiloxanes (or organosilanes). In general, the homogeneous platinum catalyst is typically not recovered and remains as an impurity in non-volatile products of the reaction. The focus of this study is on recovery of Pt from polymeric siloxanes that have been chemically functionalized through hydrosilylation using Karstedt's platinum catalyst. The products of this reaction are visibly colored light yellow to dark black due to the platinum metal impurity that results from decomposition of residual catalyst over time. Fixed-bed adsorption technology using thiol-functional silica adsorbent has been shown to be an effective technique for Pt recovery from organosilicon compounds synthesized via hydrosilylation. Here we demonstrate that colloidal thiol-functional silica particles can be used to scavenge residual platinum. The thiol-silica particles (~2 microns in diameter) were synthesized from (3-mercaptopropyl)trimethoxysilane. After washing and drying the product, the resulting colloidal silica particles were in the form of a dry powder that can be dispersed into the product of hydrosilylation. The silica particles are large enough in diameter to easily be separated from the product by centrifugation. This was an effective method for platinum product color removal since the products showed much reduced color and lower platinum metal content.

638458 The Effect of B-Site Composition in Perovskites for Electrochemical CO2 Reduction in Solid Oxide Electrolysis Cells

Abstract:

Solid oxide electrolysis cells (SOECs) are promising systems for electrochemically converting CO2 to CO, which can then be converted into synthetic fuel or other chemicals using existing technologies. Ni-based SOEC electrodes have promising electrochemical activities, but they have limited longer-term performance due to poor redox stability and carbon deposition. Mixed ionic-electronic conducting (MIEC) oxides have been explored as alternatives to minimize these adverse effects due to their favorable redox properties. However, the electrochemical activities of MIECs are inferior to those of Ni-based cathodes. Therefore, improving electrochemical performance of MIEC-based SOEC cathodes requires an understanding of the factors that govern their activity.

This study investigates the effect of B-site cation composition on the electrochemical performance of CO2 reduction in SOECs. To accomplish this, SOECs were synthesized with consistent LSGM electrolyte and LSGM-LSCF anode electrocatalysts, while various 3d transition metal cations of La-based perovskites (LaBO3; B=Fe, Co, Ni, and Cr) were used as cathode electrocatalysts. Linear sweep voltammetry profiles (I-V curves) were used to compare the electrochemical performance of SOECs as a function of perovskite composition at the cathode. We find a significant dependence on the nature of the B-site cation. These studies can be used as a guidance toward improving the performance of MIEC-based SOECs.

638274 Organic Polymers in Electrochemical Fabrication of Superconducting Interconnects

Abstract:

Rashad Williams, Mississippi State University

Faculty advisor: Dr. Qiang Huang, University of Alabama

The world is always looking for ways to improve computers or to process data faster. Most traditional computers operate information as 0 or 1. Interconnect is a network of conducting wires in an integrated circuit, which computer uses to transmit the digital information between devices. Quantum computer is a completely new device that can perform quantum computations such as superpositions and entanglement. It is no longer based on the two discrete digits of 0 and 1 like the traditional computers. One of the most promising quantum computers is operated at extremely low temperatures, typically under a few Kelvins. In such computers, any heat and noise from chemical and physical changes will hinder the computation performance. In order to avoid such heat generation or thermal perturbation, the interconnects need to be made of superconducting material, which conducts electricity without any resistance when the temperature is below a threshold point, or the so-called critical temperature. Under this condition, no heat will be generated due to the passage of electricity and thermal noise is not of a concern anymore. The long-term goal of this project is to develop electrochemical processes to fabricate such superconducting interconnects, and the specific goal of this REU project is to investigate polymer molecules for such fabrication processes. Polymers have been typically used as additives in electrochemical processes to fabricate the traditional interconnects, where the polymer molecules suppress the electrochemical deposition rate of conducting metals. However, such studies have never been reported for superconductors. Tests will be carried out to evaluate various candidate polymers and to identify good suppressors for superconductor deposition.

637688 Evaluation of the Productivity and Kinetics of Hercynite for Solar Thermochemical Water Splitting

Abstract:

As the world's dependence on fossil fuels is being reevaluated due to their harmful side effects, the search for promising renewable sources of energy has become increasingly important. Although there are several potential "green" sources of energy, the use of hydrogen as a fuel is particularly intriguing due to its high energy density. Solar thermochemical water splitting via redox reaction offers a two-step process in which specific metal oxide intermediates are subjected to a redox scheme, producing hydrogen from steam during oxidation. These redox active materials feature specific thermodynamic and kinetic properties, and iron aluminate ($FeAl_2O_4$), or hercynite, has been identified as a viable material for this process under both temperature-swing and isothermal conditions. In this study, three iron aluminate materials with different iron loadings (FeAl2O4, Fe1.2Al1.8O4, and Fe1.4Al1.6O4) are evaluated to compare the effects of iron content on the abilities of the materials to reduce CO_2, an analogous reaction to water-splitting. The various samples in this work are subjected to reduction at 1400 °C and oxidation at temperatures ranging from 1000-1400 °C. Kinetic analyses of each sample are performed to identify and model the kinetics of CO_2 reduction and determine the effects of iron loading. Comparisons of the CO productivities of the materials at various oxidation temperatures also quantify the role of iron loading in determining the ability of the materials to effectively reduce CO_2.

637801 Kinetics of Perfluorooctanoic Acid and Glycerol-Derived 1-3-Diethoxypropan-2-Ol Oxidation on Solid Catalysts

Abstract:

Catalytic oxidation reactions can provide economical solutions to widespread environmental concerns, such as decomposition of environmental contaminants and replacement of crude-oil-derived chemicals with renewable substitutes. This report presents kinetic studies of the ozonolysis of perfluorooctanoic acid (PFOA) over solid iron(III) oxide (Fe_2O_3) microparticles, and the selective oxidation of 1,3-diethoxypropan-2-ol (DEP) over platinum-bismuth nanoparticles supported on SBA-15 (PtBi/SBA-15). PFOA is a prototypical perfluoroalkyl substance (PFAS), a class of harmful contaminants plaguing the United States' and other nations' water systems. DEP and its selective oxidation product, diethoxypropan-2-one (DEK), are derivatives of glycerol that may lead to greener alternatives to existing CO_2 absorbents produced from petroleum. Both reactions were studied using an iterative reactor design process to maximize reactor reusability and minimize volatile component (e.g., solvent, reactant, product) losses. Ultra-performance liquid chromatography tandem mass spectrometry (UPLC-MS) was used to analyze PFOA oxidation products, while gas chromatography with a flame ionization detector (GC FID) was used to analyze DEP oxidation products. Analysis of products of PFOA ozonolysis suggests a rate dependence on temperature consistent with the conversion—and not adsorption—of PFOA on Fe_2O_3. UPLC-MS indicated that new, presently unknown molecules were formed during reaction. DEP oxidation selectively formed DEK in water, but little to no DEK formed in neat solution or in other solvents. This reaction was further analyzed by measuring the reaction rate as a function of the inlet air flowrate, temperature, reaction time, and DEP concentration. The trends from these studies suggest that 0th order behavior is easily achieved at relatively low inlet air flowrates, that the apparent activation energy is ~86 kJ mol^{-1}, and that the rate passes through a maximum with increasing initial DEP concentration up to ~0.2 M before decreasing at higher DEP concentrations. These results are consistent with previously reported studies of alcohol oxidation over supported Pt nanoparticles. Efforts to scale up the synthesis of DEK (from 3 mL to 300 mL) in order to perform physical property measurements of DEK have thus far been unsuccessful. DEK formation rates and DEP conversions are miniscule, suggesting either catalyst deactivation or poisoning by DEP, DEK, or a side product. These issues are discussed, along with opportunities for either oxidative dehydrogenation with alternate catalysts or oxidants, or direct dehydrogenation to produce DEK and hydrogen.

638130 Synthesis and Characterization of Non-Precious Bimetallic Catalysts for Fuel Cells Using Kudzu Biomass Materials

Abstract:

Fuel cells such as proton exchange membrane fuel cells (PEMFC) and anion exchange membrane fuel cells (AEMFC) are devices that generate electricity as the result of an electrochemical reaction, with the only byproduct being water. State of the art PEMFC work in an acidic environment and require the use of expensive platinum-based catalysts to expedite the electrochemical reactions happening in the anode and cathode electrodes. Specifically, the oxygen reduction reaction (ORR) occurring at the cathode electrode, necessitates high loadings of platinum catalysts to improve the efficiency of the cell. AEMFCs work in an alkaline environment which opens the possibility of using cheaper transition metals to form catalytic centers required for the ORR, which can potentially reduce the cost of the fuel cells. Typically, a non-precious metal ORR catalyst is formed by combining a metal source, a nitrogen source and a carbon source used as the support of the catalytic centers. In this work, we present the synthesis of ORR catalyst using iron, copper, cobalt and nickel as the metal source, an amino acid as the nitrogen source, and kudzu leaves as the precursor of the carbon support. Materials were characterized using the rotating ring disk electrode technique to determine their catalytic activity in alkaline electrolyte, and, in addition, BET studies were carried out to determine their surface area and pore size distribution.

637748 The Effect of N-Tol-Pyridinium Deposition Time on Copper-Mediated CO2 Reduction

Abstract:

Due to the rise of CO_2 emissions around the world, there is great interest in converting atmospheric CO_2 into useful multi-carbon products such as ethylene and ethanol. This goal can be accomplished by electrocatalytic CO_2 reduction (CO_2R) on copper electrodes. Our team has previously discovered that by using N-substituted aryl pyridinium additives, the selectivity of copper electrodes can be enhanced. As a hydrophobic organic layer is deposited onto the surface of a copper electrode, it suppresses the hydrogen evolution reaction (HER) and the formation of undesired side products such as CO and CH_4, leading to higher selectivity for multi-carbon products. Previously, the additive has been deposited *in situ* during electrochemical CO_2R and the precise film thickness has yet to be elucidated. This study is focused on analyzing the effect of additive deposition time, which may affect the film thickness, on our system's CO_2R selectivity profile. In conjunction, we are looking at the effect of deposition time on the electrochemical active surface area (ECSA). The ECSA represents the area on the copper electrode where CO_2R can occur. We suspect that a "sweet spot" deposition time may exist, in which the selectivity profile is optimized, and the resulting film thickness can be probed.

637882 Electronic Structure-Selectivity Trends for Electrocatalytic Nitrate Reduction to Ammonium on 3*d* Transition Metals

Abstract:

Nitrate is a toxic anthropogenic groundwater contaminant, accumulating in drinking water and posing a human-health hazard. Nitrogen loading and accumulation of aqueous nitrate in groundwater is of a comparable scale and pervasiveness as global CO_2 emissions. In parallel, production of ammonia (NH_3) by the thermal Haber-Bosch process accounted for 1.8% of the global energy consumption and represented the most carbon-intensive commodity of 2020.[1] This project aims to understand how Faradaic efficiency (FE) and selectivity towards ammonium (NH_4^+) during the electrochemical nitrate reduction reaction (NO3RR) is dependent on bulk, resting material electronic structure. Here we present NO3RR activity and selectivity of several transition metal foils (e.g. Cu, Ni, $Ni_{0.67}Cu_{0.33}$, Fe, and Ti), relating this information to electronic structure (work function, , and *d*-band center, E_d, relative to E_{vac}) and how it plays a critical role in proton affinity and NO3RR FE and selectivity to NH_4^+.

636851 Modeling Electrode Geometry Changes in Zinc-Based Batteries Using Comsol Multiphysics®

Abstract:

Zinc-based batteries are a major topic of interest in the search for energy storage technologies beyond lithium-ion batteries. In comparison to lithium, zinc's relative abundance, lower cost, higher volumetric capacity, and ability to be paired with aqueous electrolytes make it an attractive option for next generation batteries. However, zinc batteries' practical applications are limited because of low rechargeability. During battery charge, zinc is continuously plated onto the working electrode; during discharge, zinc is stripped from the working electrode. Over numerous battery cycles, inhomogeneous zinc deposition leads to the formation of dendrites, or small tree-branch-like growths of metal. These dendrites are responsible for zinc-based batteries' low rechargeability because they can (1) bridge the gap between the working and counter electrodes, causing the electrochemical cell to short circuit, or (2) break off at the root, forming "dead zinc" which has no electrical contact with either electrode and leads to capacity losses. An attractive approach to suppressing dendrite formation over multiple battery cycles involves changing the geometry of the working electrode to achieve uniform zinc electrodeposition. In this investigation, COMSOL Multiphysics® software was used to model the plating and stripping of zinc onto and from a zinc metal electrode in a two-electrode cell. As expected, greater plating/stripping of zinc was observed at the points on the two electrodes closest to each other where electrode current density is highest. The models generated in this study lay the groundwork for zinc electrode geometry optimization in the future, in which electrodes are designed to achieve homogeneous zinc electrodeposition over many battery cycles.

638436 Simulation of Colloidal Diffusiophoresis in a Microfluidic Flow Cell

Abstract:

Diffusiophoresis is the advective transport of colloids in a fluid driven by a concentration gradient of solute. For charged colloids in an electrolyte gradient, diffusiophoresis results from solute osmotic pressure gradients and the electric field induced by relative motion of the charged colloid and electrolyte ions. Diffusiophoretic fluxes can exceed colloidal diffusion fluxes, so that diffusiophoresis may be the predominant transport mechanism for colloids whenever there is an electrolyte concentration gradient, and diffusiophoresis may provide a mechanism for separation of colloids based on charge. The goal of this project was to simulate colloidal diffusiophoresis in an imposed laminar flow in a microfluidic channel, as microfluidics are well suited to conducting diffusiophoresis experiments. The coupled advective diffusion equations for the solute and the colloid were solved using the COMSOL finite element simulation package. Simulations addressed how the microfluidic channel influenced experimental measurements. Particular attention was paid to the formation of steady-state solute gradients and colloid concentration profiles near the channel inlets, the variation of colloid concentration across the depth of the channel, and the possible variation of colloid diffusiophoretic mobility in response to local electrolyte concentration. Simulation results replicated key experimental observations and indicated 1) that sharp solute gradients near the inlets cause an abrupt accumulation of colloids, 2) that colloid concentration variation across the channel depth had negligible effects, and 3) that colloid mobility variation with electrolyte concentration had a small effect on the steady state colloid concentration profile.

636520 Simulating the Resonant Soft X-Ray Scattering of Realistic Multiblock Copolymer Morphologies

Abstract:

Block copolymers are a class of soft materials that can self-assemble into mesocrystals, with applications ranging from polymer electrolyte membranes for batteries to drug-eluting coatings for biomedical devices. The detailed structure of multiblock copolymers is challenging to characterize due to the lack of chemical specificity of conventional techniques such as hard X-ray scattering. A technique called resonant soft X-ray scattering (RSoXS) leverages the unique absorption spectra of different chemical components in multi-material systems to produce scattering patterns with chemical sensitivity. Block copolymers that phase-separate into periodic structures are ideal model systems to study with RSoXS simulation tools because their scattering patterns are simple to predict. We use a resonant soft X-ray scattering simulation software (Cy-RSoXS) to study the complex structure of multiblock copolymers. We have developed Python-based scripts to create voxelized simulations of cylinders on a hexagonal lattice, a common block copolymer structure. We then implemented algorithms to simulate real samples in which the cylinders have varying radii and spacing, or the interface between different blocks exhibits a compositional gradient. The effect of these realistic features manifest into changes in the scattering pattern, such as changes in scattering intensity, peak location and peak shape. Cy-RSoXS gives us the capability to study these effects, understand phase purity, and develop more robust process-structure-property relationships. The simulation platform we have developed will enable researchers studying block copolymers with RSoXS to more accurately predict the energy-dependent scattering of realistic multiblock copolymers.

638454 Modeling and Design of Carbon Molecular Sieve Membranes for CO_2 Capture from Combined Cycle Flue Gas

Abstract:

Chief among concerns in the emergence of environmental accountability is the magnitude of pollution stemming from energy generation and industrial use of fossil fuels. As such, carbon capture measures helping to curtail emissions of greenhouse gases, mostly CO_2, have proven to be vital. Though many standing practices are effective at decreasing CO_2 levels, a more novel but also promising form of Carbon Capture and Storage (CCS) technology is that of membrane separations, specifically of Carbon Molecular Sieve (CMS) varieties. These membranes show strong ability to separate challenging pairs of gases with highly selective permeabilities, with the added potential of being more cost effective and energy efficient than traditional CCS systems.

In this work, a CMS system is modeled based on experimental data from the literature to separate an NGCC (Natural Gas Combined Cycle) flue gas stream taken from a U.S. Department of Energy benchmark amine-scrubbing CCS system. The simulated model shows the membrane's ability to separate CO_2 from the mixture, with expected decreases in operating costs. Simulation results also indicate that the membrane setup can achieve similar results to the amine scrubbing unit for CO_2 separation, but the feasibility of implementation on this scale is yet to be determined. A preliminary economic analysis of the membrane was then undertaken to provide a costing estimate for the modeled system. The decrease in energy consumption and high degree of separation parameters associated with this setup suggest that CMS membranes could be viable alternatives to conventional CCS systems.

638483 CFD Investigations of Industrial Mixing in the Transition Region

Abstract:

Mixing is commonly performed in the manufacturing and processing industry. To optimize the mixing efficiency, understanding the process details inside the tank is a crucial step. Industrial mixing is typically studied and modeled in the laminar or turbulent regimes, as the equations governing these regimes are well understood. Due to the complexity of transition mixing, there has been minimal modeling done despite transition mixing being a common working regime for many industrial processes. The lack of accurate transition mixing models hinders industries in optimizing their processes and inhibits the investigation on the behavior of the transition regime.

Computational Fluid Dynamics (CFD) simulations are used to analyze the properties and behaviors of fluid domains within a mixing tank. ANSYS Fluent, a commercial CFD software package, is used in modeling various physical systems to analyze velocity profiles, streamlines, volume fraction distribution, and numerous other fluid and flow properties. The goals of this research are to investigate the behavior of a non-Newtonian Carbopol being mixed in the transition flow regime using CFD, and to adapt these computer-generated models to physical systems to increase the accuracy of the transition mixing models. The viscosity of a non-Newtonian fluid depends on shear rate, which results in different Reynolds numbers throughout a tank, since the Reynolds number is a function of viscosity.

Three different Reynolds numbers within the transition regime were modeled: 124, 271/297, and 530. Two sets of models were generated, each with the Reynolds numbers listed previously: (i) with changing impeller speed and (ii) with changing viscosity profile. This also allows for an investigation on the flow field's behavior with changing viscosity versus changing impeller speed despite having the same Reynolds number. Using data from actual mixing systems, models of the systems were generated, analyzed, and modified using ANSYS Fluent to accurately represent the physical systems. Data in the form of pictures, videos, and torque readings will be used to compare the real system to the models. This investigation will lead to more accurate transition mixing models, a better understanding of the effect of different variables on flow domains with the same Reynolds number, as well as a deeper understanding of the transition regime.

638303 CFD and PBM of Latex Particle Aggregation in Mixing Tanks

Abstract:

This project focuses on modeling both the mixing properties and particle aggregation involved in an industrial emulsion manufacturing process. The system consists of a discrete phase of latex particles dispersed in a continuous aqueous phase. During the mixing process, small latex particles aggregate into larger and more uniform particles, thereby creating a particle size distribution (PSD). Particle aggregation and colloidal interactions are governed by the Derjaguin, Landau, Verwey, Overbeek (DLVO) theory. Brownian motion leads to the perikinetic aggregation of latex particles. Computational Fluid Dynamics (CFD) along with Population Balance Modeling (PBM) are used to track the PSD of latex particles, allowing for simulations and computational models of emulsion mixing processes, such as that of Latex, to be developed. The aim of this project is to better understand the complexities associated with colloidal dispersions and emulsion aggregation in different sized mixing tanks especially when trying to scale-up such mixing processes. Through the examination of velocity profiles and the change in PSD over time, two tanks of different sizes can be modeled and compared. CFD-PBM simulations of latex particle aggregation in water were carried out for a Pilot Plant tank and a larger Manufacturing tank using a total of three user-defined functions (UDF's). A UDF is an additional piece of code inserted in the ANSYS program to enable more complex calculations to be performed in conjunction with the CFD analysis. The first UDF, based off the DLVO theory, is used to calculate particle aggregation rates. The second UDF, a velocity UDF, is used to specify the different velocities experienced by the impeller during different stages of the mixing process. The third UDF, a viscosity UDF, is used to determine the change in viscosity of the mixture over the entire process. Our goal is to be able to formulate a scenario where the pilot tank can be used to better understand the manufacturing system.

637874 Optimal Therapy Design Using Artificial Neural Network Surrogate Models of Fluid and Solute Transport in Tumors

Abstract:

The application of quantitative and formal methods in cancer research can enable a fundamental understanding of the limitations of conventional therapies, uncover novel treatment strategies, and help optimize therapy for individual patient outcomes. Particularly, normalization of the tumor microenvironment (TME) has been proposed to improve the delivery of oxygen, antibodies, and nanomedicines to solid tumors and in turn improve treatment efficacy. A physiological transport framework based on a mechanistic model has been established to simulate the fluid and macromolecule transport in a tumor (Baxter et al). This model was used to solve parameter estimation problems to fit the model against *in vivo* experimental data using deterministic global optimization. However, it is found that the parameter estimation problems are computationally expensive to achieve global optimality using currently developed bounding routines for the mechanistic model. In this work, we propose to use artificial neural networks (ANN) to establish surrogate models for solving the parameter estimation problems to global optimality. In addition, inequality constraints are added on the feasible set of these problems based on the interstitial fluid pressure data (Martin et al). Furthermore, we also propose to use the ANN surrogates to enable the solution of a simultaneous dosage and macromolecule design problem for better patient outcomes. The use of ANN models has been confirmed to provide accurate solutions for solving challenging global optimization problems with substantially accelerated speed. The experimentally validated computational modeling approach enables TME-normalizing therapy design and anticancer drug design as well as provides insight into how in silico modeling approaches aid in predicting dose response in preclinical studies.

References:

Baxter LT, Jain RK. Transport of fluid and macromolecules in tumors. I. Role of interstitial pressure and convection. Microvascular research 1989;37(1):77–104.

Baxter LT, Jain RK. Transport of fluid and macromolecules in tumors. II. Role of heterogeneous perfusion and lymphatics. Microvascular research 1990;40(2):246–263.

Baxter LT, Jain RK. Transport of fluid and macromolecules in tumors: III. Role of binding and metabolism. Microvascular research 1991;41(1):5–23.

Baxter LT, Jain RK. Transport of fluid and macromolecules in tumors. IV. A microscopic model of the perivascular distribution. Microvascular research 1991;41(2):252–272.

Martin JD, Panagi M, Wang C, Khan TT, Martin MR, Voutouri C, et al. Dexamethasone Increases Cisplatin-Loaded Nanocarrier Delivery and Efficacy in Metastatic Breast Cancer by Normalizing the Tumor Microenvironment. ACS Nano 2019 jun;13(6):6396–6408.

638005 Analysis of Velocity and Pressure Gradients to Predict Stroke Using Carotid Artery Simulations

Abstract:

The purpose of this poster is to demonstrate the combination of computational fluid dynamics (CFD) and medicine to better characterize the early signs of stroke and aid in medical decisions. Stroke is one of the leading causes of death in the world. In the US alone, strokes account for one in twenty deaths, out of which 87% are identified as ischemic strokes. Reliable prediction of stroke is inherently challenging but is important to prevent long-term adverse effects in patients who require surgery and to avoid unnecessary surgeries in other patients. This project aims to correctly predict which patients will benefit from surgery by using CFD to analyze flow patterns in patients' carotid arteries, with the ultimate goal of creating a predictive tool that helps physicians make informed decisions regarding stroke patients. CFD is a computational technique that uses finite element analysis (FEA) to discretize a fluid domain and determine the flow characteristics and pressure gradients. The combination of CFD and a patient's specific carotid artery geometry can provide greater insight into the flow patterns of stroke and non-stroke patients. Therefore, the goal of the project is to create a patient specific CFD model to predict stroke. Computerized tomography angiography (CTA) scans from a patient's left and right carotid artery are used to generate a 3D geometry to import into a commercial CFD simulation software. In collaboration with the University of Rochester, Department of Neurosurgery, the analysis of 48 different pairs of carotid arteries (26 stroke and 22 non-stroke) has been completed. A laminar viscous flow model was utilized to model blood flow in the patient's carotid and each patient's specific peak systolic velocities (PSV) were used as boundary conditions. Results showed that stroke patients have irregular velocity gradients and high velocity values in the ICA's sinus. Recent work has shown that stroke patients have small regions of high-pressure gradients around the walls of the carotid which could be an indicator of potential stroke.

638014 Effect of Nanoporous Confinement on the Compressibility of Water: A Molecular Simulation Study

Abstract:

The thermodynamic properties of fluids confined in nanopores have been shown to differ from those of the bulk (non-confined) fluid. Thermal expansion, freezing point, density, and transport coefficients are all examples of properties affected by nanoconfinement. Of particular interest is the effect of nanoconfinement on a fluid's elastic properties, which determine the speed of elastic wave propagation through the material. When a porous material is saturated with a fluid, its elasticity is measurably changed. This makes ultrasonic wave propagation an important technique in geophysical characterization to extract information about the presence of fluids in geological porous media. Within the last five years, Dr. Gor's group has examined the thermodynamic properties of nanoconfined fluids, and in particular compressibility, with a main focus on cryogenic liquids (nitrogen, argon, methane). In this research, the methods used in Dr. Gor's group are applied to a study of the compressibility of confined water. This research aims to study the effects of the pore size and the external pressure on the compressibility of water through molecular dynamics and Monte Carlo simulations. An interest in water is justified by the substance's ubiquity in nature. The effects being studied are currently underexplored, and the results contribute to a better understanding of wave propagation in water-saturated nanoporous media.

638613 Modeling Exothermic Reactions for Pharmaceutical Intermediates in a Tubular Reactor with a Static Mixer Using Comsol Multiphysics

Abstract:

Medicines for All utilizes unique chemistry applications to develop cost effective processes to positively impact global health, and recently has partnered with Phlow Corporation to increase the pharmaceutical manufacturing capabilities in the United States to combat possible supply chain disruptions abroad. As a continuation of these objectives, a COMSOL Multiphysics® simulation was developed for modeling highly exothermic reactions carried out in tubular reactors. The COMSOL Multiphysics® model incorporated the fluid dynamics, energy transfer, species transport, and reaction chemistry through both a simplified 2D axisymmetric geometry as well as a representative 3D reactor with a static mixing element and predicted the resulting temperature and concentration profiles. In this way, the effect of the reactor design and processing conditions on "hot-spot" formation and its consequent effect on the reaction performance can be modeled in locations where temperature probes cannot practically be placed. The end goal of this model is to increase understanding for continuous processes with exothermic reactions in tubular reactors, which in turn can aid in experimental design as well as making informed scale-up decisions.

638342 Development of a Computational Tool for Solvent Recovery from Process Waste Streams

Abstract:

The global chemical industry is rapidly growing as production increases to meet the growing demand. In the United States, the amount of solvent waste from processes in the chemical industry is increasing annually, shown by the US EPA (Environmental Protection Agency) Toxic Release Inventory database. A widely used disposal method of solvent waste is incineration, but it is unfavorable due to its negative impact on the environment by releasing pollutants and greenhouse gases. In an aim to improve the overall greenness and sustainability of chemical processes, a solvent recovery computational tool is proposed.

In this work, the programs MATLAB and GAMS (General Algebraic Modeling Software) are used together to build the solvent recovery tool. The role of MATLAB is for ease of use on the user's side, therefore it is used to develop a Graphical User Interface (GUI). This GUI will allow a user to specify input parameters for a specific process or solvent waste stream and receive the solution. The role of GAMS is to perform an optimization using a superstructure-based approach. The optimization method is a superstructure-based approach and is presented as a pathway of separation units to achieve the desired separation of the solvent waste stream, while also providing economic and environmental outputs. This superstructure consists of mathematical models of separation technologies, made of mass, energy balance, design and cost equations. The superstructure considers desired recovery or purity and the process is analyzed as a multi-objective optimization problem. Using specified desired outputs, the GAMS code will optimize the recovery of solvent while minimizing economic and environmental impacts. Using GAMS as the computational tool, a case study for the recycling of Polyethylene terephthalate (PET) using ethyl benzoate as the solvent to provide proof of concept. This case study will serve as a strong example of the broad impacts that the implementation of solvent recovery can have across multiple industries.

638633 Automatic Process Synthesis for Olefin Production from Shale Gas

Abstract:

Automatic Process Synthesis for Olefin Production from Shale Gas

Arsh Bhatia, Zewei Chen, Rakesh Agrawal

Davidson School of Chemical Engineering, Purdue University, West Lafayette, IN 47906

As the world transitions toward sustainable energy sources, shale gas provides a strong option as a carbon source for this period. Shale gas is rich in natural gas and is a good source for olefin production which is used in manufacturing plastic, rubber etc. The current shale gas manufacturing processes use a sequence of front-end separation, natural gas activation and back-end separation. This has led to repeated separations which drive up costs and energy consumption. As a result, large amount of shale gas at remote areas are directly flared, leading to increased wastage of shale resources and carbon emission. Recently, developments by Chen et al. [2] have shown that eliminating some of the front end separation and postponing them for after NGL activation reduces costs and energy consumption, leading to a more intensified process design.

Chen et al.'s work, although provided a systematic procedure to rearrange the process sequence, still highly rely on researchers and practitioners' engineering intuition. In this study we investigate the automation of the synthesis procedure using mathematical programming and relying on the optimization framework to decide the optimal process arrangement. By creating an algorithm that can construct the proper search space and sort through configurations based on deciding factors such as cost and energy demand, we are able to identify unique alternatives which are more efficient.

We consider a small-scale liquid production plant from natural gas liquid in shale gas consisting of a dehydrogenation reactor and an oligomerization reactor. The algorithm considers a reaction-separation network wherein all reactions and separations are nodes and all submixture streams are edges. The algorithm can synthesize the conventional configuration, the configurations in Chen et al's work as well as several other attractive process options.

References

[1] Z. Chen and R. Agrawal, "Alternative Processing Sequence for Process Simplification, Cost Reduction, and Enhanced Light Olefin Recovery from Shale Gas," *ACS Sustain. Chem. Eng.*, 2021, doi: 10.1021/acssuschemeng.1c05091.

[2] Z. Chen, Y. Li, W. P. Oladipupo, E. A. Rodriguez Gil, G. Sawyer, and R. Agrawal, "Alternative ordering of process hierarchy for more efficient and cost-effective valorization of shale resources," *Cell Reports Phys. Sci.*, p. 100581, 2021, doi: 10.1016/j.xcrp.2021.100581.

637643 Extracting Anisotropy Strength and Interfacial Free Energy of Al-Zr Alloy Under Rapid Cooling Conditions Using Molecular Dynamics Simulations

Abstract:

Additive manufacturing is one of many growing industrial techniques with the benefit of creating new and unique materials, designs, and especially alloys. Dendrites are branching microstructural defects that will always form during the manufacturing of these alloys. Understanding and controlling the solidification to limit the formation of these dendrites is in high demand in the additive manufacturing field. The solid-liquid interfacial free energy and its associated anisotropy are key quantities to determine the microstructure formation of these dendrites. Other research has shown the addition of 0.7wt% Zr to an Aluminum Alloy AA5083 significantly increased the mechanical properties of the alloy. It is possible this is due to the dendrite formation on the atomic level and thus an Al-1at%Zr system is analyzed for this experiment. Molecular Dynamics (MD) simulations implemented in the Large-scale Atomic/Molecular Massively Parallel Simulator (LAMMPS) were used for this atomistic modeling. Using an Embedded Atom Method (EAM) potential, a Nose-Hoover thermostat and barostat, and a time step of 1 fs, a large, thin structure of Al-1%Zr was created with a solid center and two liquid ends. The system was allowed to equilibrate under an NPT ensemble at the melting/undercooling temperature and two rough, quasi-1-dimensional solid-liquid surfaces were created. The Capillary Fluctuation Method (CFM) was then utilized to analyze the solid-liquid interface. CFM relates the Fourier Amplitude of the rough surface to the wavenumber, k, and with it, the interfacial stiffness, S, is calculated for three different crystallographic orientations. Through a cubic harmonic expansion, the interfacial free energy, γ, as well as the anisotropic parameters ε_1 and ε_2 are then determined. With this, the interfacial free energy and the anisotropic parameters are analyzed at various compositions to find any correlation.

Acknowledgements: The work is supported by ARL Grant No. W911NF-2020032 and used the Extreme Science and Engineering Discovery Environment (XSEDE) TACC at the stampede2 through allocation [TGDMR140131]. This work utilized resources from the University of Colorado Boulder Research Computing Group, which is supported by the National Science Foundation (awards ACI-1532235 and ACI-1532236), the University of Colorado Boulder, and Colorado.

638331 Modeling and Optimization of a Membrane Reactor for the Steam Methane Reforming Process

Abstract:

Hydrogen has had broad usage in both the chemical and oil industries. However, its use is projected to expand into the transportation and energy markets in the future. Thus, there is a critical need to look further into methods of hydrogen production in order to economically meet those new, emerging demands. The objective in this research is to develop a model for a membrane reactor (MR) for use in the Steam Methane Reforming (SMR) process and to perform an economic analysis for optimization of its implementation into a conventional plant. Initially, the model is developed in Aspen Custom Modeler (ACM) and validated against the literature to compare the performance of three different permeation models of hydrogen through a Palladium (Pd)-based membrane. After validating the permeation models, the one that yields the lowest errors against the validation literature (4%) is chosen to replace the conventional reformer in the Aspen Plus SMR plant simulation.

With the modified design, an economic optimization is performed seeking hydrogen production maximization, using as decision variables the furnace and water-gas shift reactor temperatures, while respecting the operational limits of the conventional plant. In addition, the optimization problem also considers pollutant contamination constraints, regarding the concentration in the pressure swing adsorption tail gas. The optimization results in no changes to the conventional plant temperatures, minimizing operational impacts. The novel membrane reactor design generates reduced equipment footprint, but increases equipment purchase costs by $80,000 and operating costs by $188,000 annually. Conversely, there is an increase in hydrogen production of 1,250 kg/h, which generates a yearly increase of $22 million assuming an average hydrogen sales price of $2/kg. Thus, the developed model corresponds to a promising application of a novel, intensified design for economic hydrogen production.

637624 Mathematical Models to Compare Shale Gas Utilization Processes.

Abstract:

The migration towards more renewable sources of energy with the intention of controlling and reducing fossil fuel emissions has become the focus of many. Because of this, there is an existing demand to develop more environmentally friendly methods to produce energy. Recent studies have shown that shale gas is a comparatively cleaner alternative to other fossil fuel sources since it has lower concentrations of sulfides making it an attractive alternative to help transition into cleaner technologies. Shale gas production in recent years has grown due to advances in drilling like horizontal fracking methods which has led to a revolution and has created opportunities for the development of multiple shale utilization technologies. In order to find the most efficient and economically viable hydrocarbon conversion technologies, various technology alternatives, feedstock, and products need to be evaluated for different feedstock compositions and product portfolios. In this work we are working towards creating a superstructure framework which will enable a rapid screening of various candidate technologies. Evaluating in silico would show the benefits of candidate process technologies, to determine future direction of research by setting performance targets for material and unit model design. We first analyze two shale gas processing alternatives: the conventional alternative for shale gas processing and the proposed CISTAR alternative. The conventional alternative consists of thermal cracking units and large scale refinement facilities which makes the process energy and capital intensive. CISTAR is trying to use catalytic dehydrogenation and oligomerization in addition to membrane based separation systems to reduce energy requirement and improve scalability of the process. We converted these process flowsheets into mathematical models defining the mass balance across units in python pyomo. After the models were validated using data from our CISTAR collaborators a sensitivity analysis was carried out varying different values for conversion factors in each reactor. As an ongoing work we are exploring more complex unit models by taking into consideration the energy balance of the process, performing a techno-economic analysis including equipment and operating costs as well as the revenue produced by the system, add other shale gas processing alternatives to our model, and finally consider different viable process configurations in order to find the optimal set up.

637039 Modeling of Solar and Wind Power Plants in West Virginia Using System Advisor Model (SAM)

Abstract:

West Virginia is currently highly dependent on fossil fuels for its energy production and consumption. However, the state presents a strong capability for wind and solar energy systems. To continue to expand renewable energy infrastructure in West Virginia, the feasibility of these systems must be studied. This project was conceptualized to study the System Advisor Model (SAM) capabilities in modeling renewable energy systems and conducting feasibility studies. In particular, the objective in this project is to model renewable energy generation for solar and wind power plants in West Virginia using SAM to provide a foundation for such feasibility studies.

In this study, six wind energy plants located in West Virginia and one solar energy plant near the border of Maryland and West Virginia are modeled using SAM. Relevant weather information from the NREL Wind Toolkit or the National Solar Radiation Database and plant specifications are used to simulate plant operations in SAM. To check the accuracy of the power generation models, monthly power generations (in MWh) are compared against the Energy Information Administration's (EIA's) reported monthly power generation for each plant, and average monthly errors were calculated. After analyzing the developed SAM models for all of the wind and solar power plants studied, the results indicate that the SAM models produce consistent trends and relatively low errors (23% on average) when compared to EIA real-time monthly energy generation data for a given year. By modeling renewable energy systems with SAM, feasibility of these systems can be determined with regard to performance and financial considerations. In particular, due to the accuracy of these produced SAM models, such models could be combined with fossil fuel power plants to develop a real-time optimization framework for the local energy grid and provide realistic ramping rates for advanced control studies of fossil fuel power plants.

638171 Towards Practical Matrix Completion Methods for Transition State Theory

Abstract:

Variational transition state theory (VTST) overcomes several of the shortcomings of the potential energy-based transition state theory (TST) in describing chemical reaction mechanisms and rates. However, because of its sheer computational cost resulting from prohibitive quantum mechanical hessians and entropy estimations, VTST is yet to be widely accepted compared to conventional TST in computational kinetics studies. In this work, we aim to reduce such costs by demonstrating that matrix completion algorithms, which exploit some underlying matrix structure in order to find missing elements, can be used to recover the frequencies of modes orthogonal to the minimum energy path (MEP). To this end, we not only develop the Harmonic Variety-Based Matrix Completion (HVMC) algorithm but also use a machine learning approach to optimize the sampling of hessians along the MEP for maximum accuracy.

We demonstrate proof-of-concept of HVMC with the MEPs of several reactions. We find that HVMC recovers the missing frequencies to high fidelity even with most elements randomly missing and that sampling requirements decrease as system size increases. Furthermore, we show that column-based sampling methods still recover the frequencies with low error, and we can augment the algorithm to take advantage of apparent row-based polynomial structures for even greater performance. With this success, our future work will therefore focus on furthering the practicality of HVMC which includes utilizing gradient-based methods to compute initial approximate eigenvalues, computing cheap eigenmodes from molecular mechanics hessians, and augmenting HVMC to handle noisy data

638451 Open Catalyst Project: Finding Similarity Among Adslabs

Abstract:

The Open Catalyst Project (OCP) is a collaborative research project between Facebook Artificial Intelligence Research and Carnegie Mellon University's Department of Chemical Engineering. Currently, an open challenge in the development of renewable energy is finding low-cost catalysts to drive reactions at high rates. The use of AI or machine learning may provide a method to efficiently approximate specific catalyst characteristics, leading to new approaches in finding effective catalysts. This research project focuses on finding similarity between adsorbate-surface configurations, i.e. adslabs. An adslab include information of the surface, such as the catalyst, the adsorbate, miller index of the surface and site type. A pretrained Dimenet++ model was applied to each adslab of interest to obtain embeddings for all atoms in the adslab. Each atom will have a 256-length vector embedding. Squared exponential kernel trick was then applied to embeddings of each pairwise atoms as an indicator of their similarity. Adslabs were treated as node in a Graph Neural Network model and the Cuthill-McKee algorithm is applied to group the similar surfaces together to ease future training and prediction. Such similarity among adslabs can be applied to improve efficiency and accuracy in future model training, such as Gaussian Processes during active learning. Fewer parent DFT calls are need and accuracy of the initial stages would be greatly improved.

638438 Prediction of Theoretical Carbide Crystal Structures Using Computational Modeling

Abstract:

Transition metal carbides (TMC) are promising catalysts due to sharing similar electronic properties with noble metals. However, TMCs exist in many different crystallographic phases, making syntheses that target specific phases a challenge.[1] Combinatorics in chemical and configurational spaces allows for discovery of new mono and bimetallic carbide phases; however, performing exhaustive Density Functional Theory (DFT) calculations to compute the properties of all possible phases, and determine ideal synthesis conditions, is computationally expensive. Machine Learning (ML) models, trained with quantum calculations, seek to accelerate this process through rapid and accurate prediction of material properties.

Here, we propose a computational workflow to predict the synthesizability of theoretically proposed Molybdenum carbide structures, specifically MoC and Mo_2C, where a combination of DFT and ML techniques are employed to accelerate the material discovery process. ML models are trained on DFT calculations of commonly reported experimental molybdenum carbide structures using AMPTorch, an atomistic ML package.[2] Structures are optimized with Particle Swarm Optimization (PSO), while constraining the crystallographic prototype of the structure. The approach is validated by comparing predicted model output to DFT calculations, after which the models are employed to predict the synthesizability of over 700 proposed MoC and Mo_2C structures, with promising candidate structures validated by additional DFT calculations. This work paves the way for rapid computational screening of the vast structural space of mono and bimetallic transition metal carbides to identify new synthesizable materials with ideal electronic properties.

[1]Shrestha, A.; Gao, X.; Hicks, J. C; Paolucci, C. *Chemistry of Materials* **2021**, *33*(12).

[2]Shuaibi, M.; Sivakumar, S.; Chen, R. Q.; Ulissi, Z. W. *Machine Learning: Science and Technology* **2021**, *2*(2), 25007.

638446 Renewable Energy Data Mapping and Integration for the P J M Region

Abstract:

The PJM regional transmission organization oversees the movement of electricity to utilities throughout the Pennsylvania - New Jersey - Maryland metropolitan area of the Eastern U.S. The growing emergence of alternative energy sectors requires location-specific weather and geological data to be studied in the optimization of these energy systems, which is necessary in the greater response to global environmental issues. The goal of this research is to create a seamless software infrastructure based on MATLAB that continuously downloads and maps load and weather data to a centralized Process Information database; this data may be used in the future to propose further integration of renewable resources in the Eastern US based on predicted fluctuations in weather patterns throughout the day, with a focus on increasing renewable resources use in the region.

To reach this goal, a MATLAB script was created to access PJM data for load, wind, and solar generation and forecast in five-minute intervals through web application programming interface calls. This script was constructed to simultaneously write data to dataset-specific PI elements (having attributes in time, location, and value) in a local PI database through server connections using the PI-Software Development Kit Library. From this database, the data can be mapped and studied on a single trend.

The MATLAB-PI connection system was found to be successful in running independently and writing data to each of the respective elements in the PI database. Results for different profiles of wind and solar generation will be mapped and discussed against load profiles in the PJM region. Continued research will aim to analyze these resources on their power generation and feasibility using time series analysis and may recognize a single renewable resource as most feasible for the PJM region based on current locational information such as existence of electric systems or plants relevant to the resource.

637699 Improve Detection of Pulmonary Hypertension Using Binary Classification

Abstract:

Pulmonary hypertension(PH) is a type of high blood pressure that affects the arteries in the lungs and the right side of the heart. It is evidenced by an increased pressure in the main pulmonary artery. Doppler echocardiograms have been used to non-invasively detect PH by tracing the images to obtain the maximum velocity of blood flow into the right ventricle. This maximum velocity(Vmax) is used to calculate the right ventricular systolic pressure using the Bernoulli equation which will enable clinicians to determine if a patient has PH. Usually, these images are traced by hand by a clinician which might lead to an overestimation or underestimation of Vmax. In addition to this, the nature of the image obtained from the scan could lead to poor tracing by a clinician. Hence, inaccurate diagnosis of the disease in a given subject. In order to address this issue, a machine could be developed and trained to accurately trace Doppler echocardiograms to obtain Vmax. The first step to this goal will be building an algorithm to classify the images into good or bad images based on previous data. Another algorithm will be used to trace the good images to determine Vmax. This research aimed at classifying images into good and bad images using a convolutional neural network(CNN) and a fully connected neural network(FNN). The models were built, trained and tested on data sets obtained from Stanford hospital. A loss function was plotted for each model to analyze how well it trained. The accuracy of each model in predicting if an image was good or bad was calculated to observe its performance on unfamiliar data. CNN had a great loss function and a 35 percent accuracy as compared to a 90 percent accuracy for the FNN. Results showed that, although FNN had a higher accuracy, it had a poor loss function and performs poorly in extracting features from images. This implies that CNN is most suitable for binary classification of echocardiograms as compared to FNN, holding all other variables constant. For future studies, the CNN obtained in this study, could be optimized by using higher epochs to reduce overfitting and for higher accuracy in predicting which image is good for usage in PH diagnosis.

636942 Ufmin: A Local Optimization Algorithm for Atomic Structures Using Machine Learning Potentials

Abstract:

Simulating atomic systems in their equilibrium configurations is necessary for accurately predicting relevant physical properties, and thus is a universal challenge in computational chemical engineering research. In essence, this process of determining geometry is a nonlinear optimization problem, and conventional methods rely on density-functional theory (DFT) to compute the energy and forces of the system at each step. DFT calculations yield highly accurate results but suffer from runtimes that scale poorly with the size of the system. Thus, reducing the number of force calls required could greatly increase computational efficiency. We present here a local optimization algorithm where the force minimization is performed on a surrogate model constructed from machine-learning (ML) interatomic potentials. The Ultra-Fast Force Fields (UF3) framework is used to construct these potentials using regularized linear regression and a basis set of cubic B-spline functions. We apply our algorithm, dubbed "UFMin", to optimize the geometries of several perturbed bulk Pt structures and a perturbed heptamer island on Pt(111). A pairwise Morse potential is used as the reference potential in all cases. We find that for all bulk Pt systems tested, only three force calls are required to reduce the norm of the force to less than 1e-3 eV/A, significantly outperforming conventional methods that do not employ ML. Applying the same convergence criterion, only four force calls are required to optimize the heptamer island system. These encouraging results warrant further testing with UFMin, particularly on more complex systems and with more realistic reference potentials. We also discuss the possibility of using UFMin to accelerate nudged elastic band (NEB) calculations, which are commonly employed to identify transition state configurations and suffer from the same reliance on DFT as conventional geometry optimization.

637495 Application of Machine Learning for in-Situ Analysis of Catalyst Quality

Abstract:

Flame spray pyrolysis (FSP) provides a fast and continuous route for the preparation of nano-sized oxide-supported transition metal catalysts that play a vital role in oxidizing unburned methane at low temperature in natural gas-fired combustion engines. Determining the performance of these catalysts requires post-synthesis characterization, which hinders the ability to continuously produce high-quality catalysts. Our work incorporates machine learning and laser-induced breakdown spectroscopy (LIBS) catalyst synthesis, allowing one to tune the experimental parameters to yield the desired product selectivity. LIBS is an ideal in-situ analysis technique to monitor particle information regardless of sample phase but difficult to analyze without machine learning due to various spectral emissions mixed with fuel, precursors, and by-products.

Herein, Pd supported on Ce-Zr-Mn oxide solid solution catalysts synthesized by FSP are employed for low temperature methane oxidation. X-ray diffraction (XRD) and Raman spectroscopy were carried out to determine the chemical structure of the synthesized catalysts. LIBS, XRD, and Raman spectroscopy data were then used to obtain the support vector machine (SVM) algorithms and predict catalyst quality. Support vector classifier was used to classify the LIBS data to predict crystalline phases and oxygen vacancy percentages, whereas support vector regressor (SVR) was chosen to predict the crystalline lattice constant from the LIBS data. To evaluate the performance of support vector classifier (SVC) and SVR; LIBS, XRD, and Raman data were collected from various FSP experiments of unique state. These parameters were tuned by different precursor/gas flow rate, precursor concentration, and the precursor concentration ratios of the dopant metals. The performance of the classifiers and regression models were calculated by comparing the predicted and experimental data. Our results recorded an average accuracy of 85.86% to detect the presence of crystalline phases and 93.08% to predict low, medium, or high oxygen vacancy ranges. The SVR model was able to predict lattice constants with an average error of 0.73%. The high performances of these classifiers/model allow one to gain a perspective of the quality of the synthesizing catalysts before post-synthesis characterization, ultimately accelerating the process of producing high-quality catalysts.

638142 A Metabolic Flux and Free Energy Analysis Software for Interpreting ^{13}c, ^{2}h, ^{18}o, and ^{15}n Isotope Tracing Data

Abstract:

Quantifying metabolic fluxes offers key insights into pathway utilization, kinetics, and thermodynamics in metabolism. Stable isotope tracers, molecules labeled with a stable isotope(s), are widely used for inferring *in vivo* metabolic fluxes. Because metabolic flux analysis has emerged as the predominant method for determining fluxes for applications in biotechnology and metabolic engineering, several software packages exist to analyze metabolic activity *in silico*. However, most existing metabolic flux analysis software do not yet support evaluation of experiments with non-carbon isotopes, which can be powerful complements to ^{13}C tracers in parallel labeling experiments and reveal the turnover rates of inorganic metabolites. Moreover, current computational platforms cannot compute metabolic Gibbs free energies for thermodynamic pathway analysis. To address these challenges, we developed G-Flux, a MATLAB-based software application for computing metabolic fluxes by tracing ^{13}C, ^{2}H, ^{18}O, and ^{15}N isotopes. By integrating the backward-to-forward flux ratio, G-Flux can also compute metabolic reaction Gibbs free energies using only isotope-labeling data. Through simulations of ^{13}C and ^{2}H glucose tracing experiments, we computed metabolic flux and free energy values of reactions within the glycolysis and pentose phosphate pathways of *Escherichia coli*. Most 95% confidence intervals of the corresponding free energies determined by G-Flux were narrower than those of the published values. Using these precedingly calculated fluxes, we further simulated mass isotopomer distributions of metabolites in *Escherichia coli* and identified [6-^{18}O] glucose to be a candidate for inferring glycolytic reversibility and [U-$^{13}C_5{}^{15}N_2$] glutamine to be a candidate for quantifying fluxes and free energies of transaminase and amino acid degradation pathways. Through quantifying metabolic rates and guiding the rational selection of optimal tracers for labeling experiments, we anticipate G-Flux to facilitate efforts in improved metabolic flux and free energy analysis.

638354 Optimization of Fructose Dehydration over Zeolite Catalysts Using Machine Learning

Abstract:

The dehydration of biomass-derived fructose is an important reaction to produce the platform chemical, 5-hydroxymethyl furfural (HMF), which could be further processed into biofuels and other green derivatives. This reaction has several side products, making the identification of selective catalysts challenging. Therefore, it has been extensively studied in experiments over various zeolite catalysts. However, no comprehensive analysis has been done to understand the relationships between the yield and rate of the reaction and the properties of different zeolite frameworks.

In this work, we developed a machine learning model to predict the yield and the rate constant of fructose dehydration to HMF and identify potentially better zeolites for this reaction. Experimental data from 25 research papers were collected. 12 features representing the catalyst and solvent properties, and reaction conditions were selected to train the model. Various regression algorithms were used to predict the two labels separately. A 5-fold cross-validation error scaled by the range of each label is used as an evaluation metric.

Random Forest regression model was found to have the lowest error. When trained on the dataset, the model had cross-validation errors of around 14 % and 20% for the prediction of yield and rate constant, respectively. Descriptors such as the TPD Ammonia, Si/Al ratio and BET surface area of the catalyst, as well as total dipole moment of the solvents were found to be the most important features to making the predictions.

The results showed that the conditions leading to a higher rate constant don't correspond to a higher yield because they promote side reactions as well. The model was used to predict the performance of potential zeolites in the IZA database. Zeolites such as Metal-Organic Zeolite (MOZ) and Linde Type L (LTL) are found to show better performances compared to the current zeolites used for this reaction. We propose that transfer learning can be combined with this model to predict the performance of reactions with similar mechanisms such as Xylose dehydration.

637001 A Data-Driven Optimization Approach and Software Toolkit for Modular Process Intensification Synthesis

Abstract:

Modular process intensification aims to develop more efficient, more environmentally friendly, and more compact technologies by combining multiple process tasks into a single equipment, maximizing driving forces, and maximizing heat and mass transfer rates. A key research question is how to systematically generate innovative process designs. To this purpose, phenomena-based process synthesis approaches have been proposed which offer the potential to re-invent unit operations by finding optimized process solutions without pre-postulation of equipment/flowsheet alternatives. Herein, phenomena are fundamental chemical process functions such as heat transfer and mass transfer, which can serve as building blocks to represent unit operations from a lower aggregated level. A major challenge in applying phenomena-based synthesis is the computational complexity due to the large-scale optimization problem and the nonlinearities introduced by modeling the first principles physical phenomena.

To address this challenge, we propose a data-driven optimization approach for phenomena-based process intensification synthesis using the Generalized Modular Representation Framework (GMF). In GMF, two types of phenomenological building blocks are used: pure heat exchange modules (HE) and mass/heat exchange modules (M/H). The mass transfer feasibility in a M/H module is characterized by the Gibbs free energy-based driving force constraints, which contribute to the key GMF representation capability to identify multifunctional separation and/or reaction tasks while also render the major mathematical complexities with logarithmic terms and multivariable polynomials. In view of this, the rectified linear units (ReLU) approach is applied to generate a data-driven model to correlate driving forces with key process variables (e.g., temperature, molar fractions) in a mixed-integer linear formulation. This offers the advantages to: (i) reduce the computational strain to identify optimal solutions, and (ii) extract simplified mathematical expressions to spotlight the synergistic relations of multifunctional phenomena. The data-driven model is then integrated with other GMF modeling constraints on mass balances, energy balances, and superstructure combination rules to synthesize modular and intensified process systems. The proposed hybrid data-driven/mechanistic GMF synthesis approach will be showcased on two case studies: (i) reactive distillation optimization to demonstrate the solution optimality and computational efficiency with respect to continuous optimization, and (ii) modular reactor network optimization for decision making with mixed-integer variables. The automation and integration of this strategy as a software toolkit to the SYNOPSIS prototype platform will also be presented.

637982 Validation of a Fast Solvent Evaluation and Selection Model for Extractive Distillation

Abstract:

Extractive distillation separates a two-component mixture of close-boiling or azeotropic components by adding a third component (entrainer/solvent) to the process. The selection of the third component is critical because it affects the column design and cost. Here, we validate a Fast Solvent Evaluation and Selection (FSES) model [1] by combining process simulation with derivative-free optimization (DFO). The model [1] employs physical properties, such as boiling point, selectivity, and molecular weight, and process properties, such as minimum reflux ratio, minimum reboiler duty, and column stages to predict solvent performance and rank the solvents.

The framework for validating the FSES model integrates a rigorous extractive distillation process simulation in the inner loop with a DFO algorithm in the outer loop iteratively. Given a separation task, the goal is to find the optimum process design and operating conditions, i.e., inlet stages, solvent ratio, column stages, distillate rates, and reflux ratios, for each solvent that the FSES model ranked. The optimum is defined as the design and operating conditions that minimize the total annualized cost (TAC). In each framework iteration, the DFO algorithm generates a set of design and operating conditions and passes them to the process simulation. Once the simulation is converged for the conditions, the total annualized cost is calculated and passed back to the DFO algorithm, starting a new iteration. This iterative process continues until the DFO algorithm no longer generates candidate design and operating conditions with lower TAC values or the maximum number of simulation runs is reached. At termination, the framework yields the minimum TAC and the optimum design and operating conditions for each solvent. Then, the candidate solvents are ranked from lowest to highest based on their minimum TACs, and these rankings are compared to those obtained from the FSES model.

We applied the validation framework to two case studies, acetone/methanol and benzene/cyclohexane separations. For the acetone/methanol separation, the FSES ranked the solvents from best to worst as dimethylsulfoxide (DMSO), water, ethylene glycol, ethanol, and 2-propanol. The framework yielded the same order. For the benzene/cyclohexane separation, the FSES model ranked the solvents from best to worst as sulfolane, furfural, dimethyl phthalate, N-methyl-2-pyrrolidine (NMP), and aniline. The framework yielded the same order, except it ranked NMP as a better solvent than dimethyl phthalate.

References

[1] Xu, S., Crump, T., Cremaschi, S., Eden, M. R., & Tula, A. K. A Short-Cut Method for Synthesis of Solvent-based Separation Processes. In preparation.

637642 Molecular Dynamics Simulations to Investigate Stability of Lysozyme Under Different Freezing Rates

Abstract:

Freezing processes are integral in drug storage and preservation, especially in the protein biotherapeutics that have complex structures which must remain unperturbed in order to ensure stability. Protein freezing results in a lessened reaction rate and decreased risk of microbial growth. An increase in hydrogen bonds after freezing inhibits flexibility and increases structural compactness. Consequences associated with freezing, such as solute rejection along the ice-water interface, induces negative impacts on stability and activity. Molecular dynamics (MD) simulations examine this phenomena in the Lysozyme protein, desired for its antibiotic and immune boosting properties . The use of MD simulations allow for the discovery of underlying causes of protein instability during freezing.

All-atom MD simulations reveal the interactions causing instability. Necessary softwares to perform simulations and analyze data include GROMACS, VMD, Pymol, and Gnuplot. Two Lysozyme proteins (PDB:1AKI) are randomly put into a simulation box (6.4nm x 6.8nm x 13 nm) with 10mM sodium phosphate buffer: pH = 7. Three "non-isothermal" models show a temperature change from 270K to 247K. One is slow freezing represented by cooling 5-6K per 490 ns. Medium and rapid freezing are represented as cooling 7-8K and 11-12K respectively over 490 ns. An ice seeding method propagates the formation of a freezing environment. Equilibration is achieved through application of 1000 conjugate-gradient minimization cycles and a 500 ps MD simulation in the NPT assembly. Temperature and pressure control is achieved using a Langevin thermostat and Nose-Hoover barostat. The equilibrated system is subjected to another MD simulation using an NVT ensemble. Simulations are subject to periodic boundary conditions, and compiled on a supercomputer cluster called HPC.

Ice seeding populates the protein buffer system to simulate freezing conditions. The system is equilibrated by an analysis of the convergence of potential energy. Root-mean-square-deviation (RMSD) evaluates the structural similarity between conformations. There are large fluctuations in early stages of freezing due to the proximity of Lysozyme to the ice-water interface, but remain stable later in the freezing process. Root-mean-square-fluctuation (RMSF) identifies the most flexible residues in Lysozyme by analyzing H-bond formation between the protein and solvent molecules as well as intramolecularly. Analyses correlate variables that affect stability in Lysozyme. These include the effects of temperature, solute concentration, and ionic strength on a number of factors dictating stability such as unfolding, aggregation, degradation, and molecular interactions. This simulation and analysis improves understanding of these correlations in Lysozyme.

638468 Design of an Open Source, Automated Drop-Casting Device to Dispense Nanoliter Droplet Arrays

Abstract:

An automated method of depositing sub-µL droplets of liquids is sought after for the manufacture of chemical sensors. By depositing polymer solutions and evaporating the solvent, thin films of functional polymers can be placed in the desired pattern to create sensing elements. Traditionally, piezoelectric ink-jetting can be used to dispense picoliter-scale droplets repeatedly until the target volume is deposited. However, piezoelectric nozzles are susceptible to clogging when used with polymer solutions. Other deposition techniques include nanoplotters which can deposit nanoliter-sized droplets by imprinting dispensed liquid droplets onto materials. This method is less desirable because less control is given over the volume of the deposited droplets and damage to the sensor is made possible by impaction. In this study, an inexpensive, automatic drop-casting device is made by modifying a 4-axis CNC machine to use with a syringe pump extruder. Marlin, an open-source 3d-printing firmware, was configured to control the XYZ positioning of the syringe pump as well as the volume of liquid dispensed. The machine successfully placed an array of 38 nanoliter droplets. The minimum droplet size is limited primarily by syringe diameter. The theoretical precision of this machine is 2.5 microns per motor micro-step in the X, Y, and Z directions. Future studies may also include hydrophobic treatments to decrease droplet adhesion to the needle tip.

638480 Improving a Power Series Approach for Heat of Vaporization Predicted By the SRK Equation

Abstract:

Equations of state like the Soave-Redlich-Kwong (SRK) equation are used to predict the thermodynamic properties of various substances. In previous research, a new adjusted temperature variable, η, was introduced to simplify calculations of equilibrium properties. The goal of this project was to continue using the adjusted temperature variable in both the SRK and Clapeyron equations to model heat of vaporization. An existing calculation algorithm for the Clapeyron equation in reduced temperature was modified to express a new variable, the adjusted heat of vaporization, as a power series in the adjusted temperature. Coefficients of the power series consisted of a constant, K_1, and a second term proportional to the acentric factor function of the SRK equation, $K_2f(\omega)$. The series using K_1 constants followed the predicted SRK heat of vaporization data to within the same multiplicative factor, which was used as an empirical adjustment. Constants up to the thirteenth series term were verified. Results matched the predicted SRK data to within 0.1 percent for all adjusted temperatures greater than 0.0418, and to within one percent for all adjusted temperatures greater than 0.0295. Work continues to replace the empirical correction with an exact calculation, to find higher order constants which converge to the SRK data, and to evaluate the constants K_2 for the terms proportional to the acentric factor function.

638680 *The Use of Comics to Educate K-12 Students on Voltaic Cells*

Abstract:

Chemical engineers are integral to the functionality of fields like the pharmaceutical industry that provides medicine to keep our communities healthy and petrochemicals to fuel our cars and homes. Through all the work that chemical engineers engage in, however, environmental sustainability is a constant factor of optimization, especially in our energy districts. On a planet plagued with the devastating consequences of climate change, chemical engineers become indispensable to not only improving these effects but also reversing them over time. However, while the work that chemical engineers do is so critical, the process to becoming one is also notoriously intimidating. Chemical engineering is one of the most difficult engineering disciplines to understand at a collegiate level, and this often discourages many from pursuing it as a degree. Not only is the content challenging, but the methods by which it is taught only exacerbates the challenge of learning the material. Monotone college lectures deter interest and convoluted textbook pages only frustrate students. However, one method that has been found to have an effective impact on helping students learn and digest class material are comics, educational narratives accompanied with well-crafted images.

One foundational concept that many chemical engineers have had trouble understanding in chemical engineering is electrochemical batteries and how they work. In the comic that we have constructed, we explain the structure, functionality, and use of different batteries that are used by many on a regular basis. In this comic, a personified flashlight explains to a struggling student the inside structure of a basic voltaic cell and what chemical reactions cause it to power objects like the flashlight itself. This conversation facilitates both a visual understanding of voltaic cells and a vicariously audible understanding as the reader is put into the position of the student 'hearing' the flashlight explain it. Using this comic to explain voltaic cells provides a colorful alternative to the tedious and perplexing teaching methods usually found in traditional classrooms and will promote interest and engagement in chemical engineering.

637696 Barriers and Assets: Understanding the Experiences of Underrepresented Undergraduate Engineering Students.

Abstract:

Abstract

Engineering and other Science, Technology, Engineering and Mathematics (STEM) student retention rates are alarming in the United States. For example, the annual graduation rate of those who enter college as STEM majors is about 50%. In addition, in 2015, less than 30% of underrepresented engineering students in the United States earned a bachelor's degree. In the location of this study, a Minority-Serving Institution (MSI), historically, only about 25-30% of undergraduate engineering students complete degrees annually. In spite of this disparity, Minority-Serving Institutions (MSIs) are leading the way in producing a large percentage of underrepresented engineering bachelor's degree holders. For example, while accounting for only 3% of all post-secondary institutions in the US, HBCUs graduate 17% of all Black engineering students in the US. Of Hispanic students pursuing a degree in engineering, at least 59% obtain their degree from an HSI. The field could stand to learn more about the experiences of underrepresented engineering students at MSIs (the largest producer of underrepresented engineers). However, much of the research about the experiences of underrepresented engineering students is in the context of Predominantly White Institutions (PWIs). Therefore, the purpose of this study, which is a part of an ongoing study, is to understand the experiences (barriers and assets) of undergraduate engineering students at a large Research I, HSI in the Southwest. The authors used a qualitative approach to understand the experiences of 10 underrepresented undergraduate engineering students (including three Chemical Engineering students) using a semi-structured interview. The authors analyzed data using Interpretive Phenomenological Analysis and were guided by Community Cultural Wealth to uncover the assets they used to succeed in engineering. The most prominent barriers that arose were professor concerns, advisor concerns, and school/life balance. The observed assets included peer interaction and support structures. Finally, the work presented here sheds light on the importance of faculty and academic advisors in students' success. Data also suggest that the most important factor in student success is engagement, specifically, their interactions with their peers, whether through formal student organizations or informal networks.

637702 Black Males in Pursuit of Advanced Engineering Degrees

Abstract:

Black males have long been underrepresented in engineering programs and the engineering workforce. Despite attempts to increase Black and other underrepresented student representation, these efforts are limited and have yet to create transformational change. There is little research that has sought to reveal and understand the factors that influence Black males to pursue engineering graduate degrees, and to discover what strengths Black males who have persisted through these programs possess. In this study we sought to understand the factors that influenced 26 Black male engineering faculty to pursue engineering graduate degrees (including 5 chemical/bioengineering faculty) and the capital that they engaged to persist. Using Cultural Capital Theory and Interpretative Phenomenological Analysis, two themes emerged, *benefits of advanced degrees (e.g. increasing and applying knowledge, and career advancement)* and *social supports (e.g. family and faculty)*. These findings may assist in creating more innovative interventions for recruiting and retaining Black male engineering graduate students.

637993 Quantifying Deliberate Practice Using Multiple Student Response Metrics on Auto-Graded Homework

Abstract:

Auto-graded homework questions with randomized numbers and content provide frequent, formative assessments for students that create big data for instructors and researchers. Online homework questions combine defined and repetitive practice, feedback on correctness, explanation of solutions, and availability of repeated assessment, which are tenets of deliberate practice. Over 800 questions and 170,000 attempts were studied from three cohorts of students (n=284) using a fully interactive online textbook, Material and Energy Balances zyBook. Question sets consisted of multiple parts where easier levels preceded harder levels. Question types included multiple choice, single numeric, and multiple numerical responses. When students were allowed an unlimited number of attempts before a fixed due date, a median correct of 94% was measured. To investigate student responses beyond correct and incorrect, multiple metrics were aggregated into a deliberate practice score. Deliberate practice scores between 0 and 8 encompassed fraction correct, modified correct, first quartile attempts before correct, and median attempts before correct at different thresholds. The deliberate practice score distinguished between question type and order. Furthermore, students disproportionately practiced questions with higher deliberate practice scores without data beyond their own correctness.

638043 Sealed Vessel Leak Test

Abstract:

The intent of this experiment is to develop a test procedure to validate leak-proof integrity of previously sealed Swagelok assembled fittings with welded tubes at one end. These vessels were previously used in support of the JT5-Alt370-IN-1 compatibility testing. Components remaining from the prior test will be assembled into vessels then submerged into a Parr Instrument Company pressure vessel. The instrument temperature and pressure conditions will be set to predetermined levels that simulate the prior test conditions. Vessel weights will be measured before and after immersion to determine if leakage occurred. Vessels will be opened and visually inspected for fluid to verify results. Once this technique has been validated the test will be conducted with the remaining IN1 vessels

LA-UR-21-26881

Copyright © American Institute of Chemical Engineers. All rights reserved.

638166 Developing a Kinesthetic Intervention for the Development of Expertise in Structural Engineering

Abstract:

Active learning is well-established as a more effective approach than passive teaching methods, such as traditional lectures. Kinesthetic learning is a relatively new subcategory of active learning, where learning is facilitated through movement or interaction with props. There is great potential for kinesthetic learning as a pedagogical technique for certain concepts. This study looks at whether a kinesthetic intervention can promote students' development of expertise in construction site hazard identification. Every year, ~1,000 construction workers in the U.S. die due to hazards and working conditions. Hazard identification is difficult to teach in undergraduate curricula as access to construction sites is limited and pictures are not representative of the real, three dimensional, world. Our study uses a 3-D printed model to supplement a hazards identification lecture. This work represents a pilot study designed to prepare the intervention for classroom use. The pilot study included two methods: (1) think-alouds to evaluate the usability of the model and (2) a small-scale control versus intervention study to gather preliminary data. Think-alouds revealed a few pieces of the model which needed to be tweaked to provide more stability for the whole model. Our pilot intervention study showed areas of improvement in the wording of the survey. Most people wrote down particular objects instead of specific hazards, and subsequently the language has been adjusted to be more specific. These results inform the design of a larger study to be conducted at Oregon State University in their Temporary Structures course. The data collected will allow us to determine whether kinesthetic learning is effective in promoting expertise development for students in construction site hazard identification that can be translatable to other disciplines, including process safety.

638000 Experimental Design for the Measurement and Analysis of Vibration-Rotation Spectra of Molecules Relevant to Extra-Terrestrial Bodies

Abstract:

High-precision spectroscopy in the near-infrared (NIR) will be central to improving future measurements in many fields, including meteorology and astronomy. For the accurate spectroscopic analysis of planetary atmospheres, including Earth's, high-precision NIR spectra for simple molecules such as acetylene, methane, water, and carbon dioxide are needed, but current data still contains myriad ambiguous and unassigned features at lower than ideal precision. Consequently, there still exists data from NASA missions such as the Cassini probe that have been unable to be analyzed due to poor laboratory references. As more space missions are initiated, notably NASA's "New Frontiers" program and the Titan Saturn System Mission, new planetary atmospheric data will be collected, and such data will require adequate high-precision laboratory data in order to make confident conclusions about those atmospheres. Recording all this data is a massive undertaking, and it will require many spectroscopists to accomplish, so cultivating interest in the field is a crucial step towards accelerating progress. Therefore, we present an experimental design that records NIR absorption spectra of simple molecules using relatively cheap materials as compared to the highest precision experiments but achieves a much higher precision than traditional off-the-shelf spectrometers traditionally used in teaching laboratories. We record several lines of acetylene as a proof of concept, and we measure Boltzmann's constant as an example for a teaching laboratory and as a benchmark for the quality of the spectrometer.

638265 A Techno-Economic Assessment of Distributed Membrane-Based Water Treatment Systems

Abstract:

Driven by an increased use of nitrogen-based fertilizers and nitrogen-fixing crops, groundwater nitrate contamination and water salinity have reached severe levels in many remote disadvantaged communities (DACs) adjacent to agricultural sites. These communities lack a centralized potable water supply and are dependent on groundwater for their drinking water. The Salinas Valley Distributed Water Treatment (SVDWT) Project was proposed as a reliable and affordable option for supplying safe drinking water to DACs where alternative measures were infeasible. In this approach, multiple wellhead water treatment systems in multiple DACs are virtually networked and operated remotely to provide economies of scale, leading to affordable operating costs. Reverse osmosis (RO) membrane systems with septic tank concentrate handling has proved to be a technically feasible means of treating groundwater in remote communities. Such systems are operationally simplistic and can remove a variety of contaminants over a wide range of nitrate and salinity levels, while meeting community water demand and regulatory water requirements. Here, we conduct a techno-economic assessment of a remotely operated wellhead water treatment system using actual expenditures and process data from a pilot plant operating in the Salinas Valley. Capital expenditures (CapEx) of the treatment system and site infrastructure (e.g. water tanks, septic tank) were calculated using recorded costs of constructing the pilot site. Operational expenditures (OpEx) were modeled using pilot plant process data. To further contextualize the cost of water treatment via a remotely monitored membrane-based system, it is compared to the modeled cost of treatment via ion exchange technology. It is demonstrated that remotely operated, distributed membrane-based water treatment systems are an economically viable solution to DACs suffering from groundwater nitrate contamination.

638691 Modified Tea Waste: A 'Greener' Wastewater Treatment Technology

Abstract:

Joyee Chakma[1‡], Md. Riad Al Rabbi Abir[1‡], Ehsanur Rahman[1], Shakhawat H. Firoz[2], Nafisa Islam[1*]

1 Department of Chemical Engineering, Bangladesh University of Engineering and Technology, Dhaka, Bangladesh

2 Department of Chemistry, Bangladesh University of Engineering and Technology, Dhaka, Bangladesh

‡ indicates presenting author

* Corresponding author email: nafisaislam@che.buet.ac.bd

Abstract:

The annual tea production of Bangladesh in 2020 was more than 79 million kg,[1] with 90% of the tea left behind as waste post-extraction and consumption.[2] Agricultural solid wastes such as those from the tea industry can be a source of readily available, environmentally friendly sorbents for wastewater treatment.[3] The adsorption efficacy of pollutants, such as methylene blue dye can be further improved by applying chemical activation agents upon tea wastes.[4] This study evaluates the dye removal efficacies of chemically modified tea waste adsorbent, using 'greener' chemicals for modification. The dye removal efficacy of five types of tea waste adsorbents were studied; unmodified tea waste (U), unmodified tea waste treated with sodium hydroxide (UNA), tea waste modified with three 'green' acids, i.e. citric acid (M-CA), oxalic acid (M-OA) and tetraacetic acid (M-EDTA). In the case of M-CA, M-OA and M-EDTA, the modified tea waste adsorbents were stabilised with sodium hydroxide (NaOH).Equilibrium and kinetic absorption studies were carried out for all five tea waste adsorbents using UV-VIS spectroscopy for methylene blue dye concentration measurement. The tea waste adsorbents showed a range of adsorption capacities (reported in units of mg dye adsorbed per g of tea waste adsorbent used) with 732.32 mg/g, 733.53 mg/g, 1002.89 mg/g adsorbent, 670 mg/g 914.742 mg/g for U, UNA, M-CA, M-OA and M-EDTA, respectively. The dye adsorption data for all the adsorbents showed Langmuir adsorption characteristics and followed pseudo-second order kinetics. The adsorbents were further characterised using Fourier Transform Infra Red Spectroscopy. The incorporation of acidic (carboxyl) groups by esterification of the unmodified tea waste, using citric acid, oxalic acid and the incorporation of complex chelating functionalities by EDTA all contributed to the increase of dye adsorption capacity of the tea waste. The stabilizing agent, NaOH did not have effect on the performance of the raw or unmodified tea waste (data not shown). The results of the chemical kinetic analysis indicate that the chemical modification accelerated the intraparticle diffusion rate. The maximum dye removal capacity using 'green' chemical activators for tea waste was higher than those found in literature, and thus these novel adsorbents can be potential low-cost 'green' solutions for wastewater treatment.

[1] BTB: Bangladesh tea board. Annual Statistical Report; 2020.

[2] https://doi.org/10.1016/j.cej.2011.04.017

[3] https://doi.org/10.1007/s11270-018-3869

[4] https://doi.org/10.1016/j.cej.2011.09.100

636911 Seasonal Impacts on Antibiotic Resistance in Oregon Wastewater Treatment Plants

Abstract:

With the overuse and disposal of antibiotics in hospitals, agriculture, and residentials, wastewater treatment plants are considered one of the main sources and reservoirs of antimicrobial resistance (AMR). To understand proliferation and persistence trends of antibiotic resistance in wastewater treatment plants, this study investigated the prevalence of eight different phenotypes of AMR and multi-drug resistance *E. coli* in 17 wastewater treatment plants across Oregon in winter and summer of 2019 and 2020. Overall, 1187 *E. coli* colonies were isolated from influent, secondary effluent, final effluent, and biosolids. 31.8% of all collected isolates exhibited resistance to at least one antibiotic, and 8.7% carried multi-drug resistance phenotypes, of which 16.2% harbored resistance to more than five classes of antibiotics. From a total 246 sampling incidences, 23 demonstrated multiple antibiotic resistance index above 0.2, suggesting high risk of antibiotic contamination. Significantly higher multiple antibiotic resistance index and prevalence of ciprofloxacin-resistant *E. coli* was observed in biosolids. Highest AMR rates amongst collected *E. coli* isolates were observed for ampicillin (18.4%) followed by streptomycin (13.8%) and tetracycline (13.6%). Additionally, resistance to trimethoprim-sulfamethoxazole showed significant associations with the occurrence of ampicillin, tetracycline, and streptomycin phenotypes ($p < 0.01$). When comparing different phenotypes of AMR between winter and summer in the influent, significantly higher prevalence of ciprofloxacin-resistant *E. coli* was observed in summer ($p < 0.01$), suggesting the seasonality of different phenotypes. Findings suggest that the removal efficiency of AMR *E. coli* in wastewater treatment plants is likely influenced by seasonal variations.

637992 Magnetic Water Filter

Abstract:

Accumulation of arsenic in the body is detrimental to health and directly leads to deadly cancerous and noncancerous diseases. Unfortunately, conventional methods are unable to effectively address selectivity and cost issues of arsenic removal in water. Previously, our lab developed engineered E. coli for As(III) bioaccumulation. To prevent secondary contamination a magnetic water filter was built with inspiration from membrane bioreactors. In this poster, I report the arsenic and bacterial removal efficiency of our filtration tank and discuss the potential of using engineered E. coli for arsenic and other metal removal.

638503 Microbial Source Tracking Method Validation to Identify Hosts of Fecal Contamination and Pathogen Presence in Source and Stored Drinking Water in Urban Kenya

Abstract:

The spread of enteric pathogens is a significant cause of disease and death in many low- and middle- income countries. These pathogens can spread from feces to hosts through multiple different pathways making sanitation interventions ineffective at completely limiting the spread of disease. Molecular source tracking (MST) assays have previously been used for the identification of host-specific feces in water and other environmental samples. For this project, human and animal fecal samples along with stored and source water samples were collected from individual homes in an urban area of Kenya. Prior to the use of the MST assays, the nucleic acid extraction methods were validated to assure the quality of the RNA and DNA extracted from the samples and their suitability for downstream applications such as qPCR. Additionally, the extraction methods were validated for use in the Taqman Array Card system which uses microfluidic assays for the detection of specific enteric pathogens. These results will aid in the overall evaluation of the sources of fecal contamination and pathogens present in the collected water samples. Additionally, a systematic literature review was conducted to examine the extent to which MST assays have been used to detect fecal contamination in low- and middle-income countries.

637972 On-Site H_2O_2 Generation for Water Reuse Treatment Systems

Abstract:

Water reuse is a current issue in many western states where widespread drought, an issue exacerbated by climate change, constitutes water shortages. H_2O_2, a green oxidant commonly used in water treatment, is added to water in batch and used to generate ·OH under UVC photolysis to remove organic and inorganic contaminants in the water. However, purchasing, transportation, and storage of bulk H_2O_2 poses a significant cost for the water reuse treatment process. This research aims at generating on-site H_2O_2 for use in water reuse treatment facilities.

Photocatalysis is a promising approach for onsite H_2O_2 generation. As in many photocatalytic processes, separation of photoholes and electrons is key to efficient H_2O_2 generation. We deposited Au nanoparticles on TiO_2 (P25) nanoparticles to form a metal/semiconductor Schottky junction and improve charge separation. P25, a powder form of TiO_2 containing both anatase and rutile phases, creates a porous film of TiO_2, allowing maximal surface area for electron transfer. We deposited gold nanoparticles onto the TiO_2 via deposition precipitation (DP), which evenly deposits Au onto the porous surface. We designed and built a millifluidic reactor to test the Au-P25 samples under UVA light (365 nm). The use of Au nanoparticles in conjunction with the TiO_2 greatly increases the generation of H_2O_2 due to the electron generated by the excited gold and transferred to the TiO_2 surface.

We explored different methods for Au DP to determine optimal loading of Au nanoparticles onto the TiO_2 surface. We also calcinated the TiO_2 films at different temperatures; calcination temperatures above 600°C began to form primarily rutile phase TiO_2, while temperatures below left primarily anatase phase. XRD supports these findings. Au-P25 samples calcinated at 650°C show substantial on-site H_2O_2 generation (>40uM), this is valuable for water treatment using H_2O_2. When samples are tested in 5% EtOH, a common electron donor that decreases incidences of recombination, there is an increase of H_2O_2 generation in rutile phase samples. However, the EtOH decreased the H_2O_2 generation rate in anatase phase samples. The proposed mechanism of H_2O_2 generation is that dissolved oxygen in the water adsorbs to the TiO_2 surface and two-electron reduction generates H_2O_2. With this mechanism, we expect EtOH would increase H_2O_2 yield on both sample types (anatase and rutile phase TiO_2). This suggests the need for further research to examine the mechanism of H_2O_2 generation on anatase and rutile phase TiO_2.

637909 Electrocatalytic Conversion of Wastewater Nitrate to Valuable Ammonium

Abstract:

Nitrate contaminated wastewater and drinking water negatively impact the environment and human health. For example, nitrate contributes to eutrophication and disruption of the nitrogen cycle as well as certain diseases and disorders such as methemoglobinemia and endocrine disorders. Conventional nitrate removal methods include ion exchange, reverse osmosis, and biological denitrification. These processes often produce concentrated A promising alternative method for nitrate removal is electrocatalytic nitrate reduction. Through this process, nitrate may be transformed either into N_2, a benign gas, or into ammonium, which enables recovery of reactive nitrogen in wastewaters for reuse as fertilizer or energy storage in nitrogen-based fuels. However, this method often requires expensive precious metals such as Pt or Pd, or metals with low electrochemical stability such as Cu.

Therefore, we investigated the use of transition metal oxynitrides (MO_xN_y) and MO_xN_y on reduced graphene oxide supports (MO_xN_y /rGO), as novel catalysts for the nitrate reduction reaction to address the lack of cost-efficient, active, and stable electrocatalysts with high selectivity to ammonium. Following a urea glass synthesis, the MO_xN_y and MO_xN_y/rGO catalysts were used as cathodes in a batch electrolysis reactor containing different initial nitrate concentrations (1 mM, 10 mM, and 50 mM) representative of those found in nitrate-containing wastewaters (polluted groundwater, textile wastewater, and industrial wastewater, respectively). Results showed increasing nitrate conversion with increasing nitrate concentration and higher conversions with MO_xN_y/rGO electrodes compared to those coated with MO_xN_y. Further electrochemical analyses with linear sweep voltammetry and electrochemical impedance spectroscopy confirmed the improved activity of MO_xN_y/rGO compared to MO_xN_y, characterized by a lower onset potential, a lower charge transfer resistance, and higher electrochemically active surface area. Furthermore, electrochemical membranes (EMs) functionalized with MO_xN_y and synthesized with reduced graphene oxide were prepared and tested for nitrate reduction. Results suggest that modular EMs for nitrate electrochemical conversion may enable recovery of reactive nitrogen as valuable ammonia.

637991 The Use of Bioremediation As a Potential Treatment for Produced Water

Abstract:

A byproduct from the production of natural gas and petroleum is produced water. Large volumes of produced water are generated annually and can be as high as 2 million gallons per unconventional well. Produced water can be reused in the fracking process, injected in storage wells after use, or treated before being released into the environment. One possible treatment option, bioremediation, uses naturally occurring bacteria in soil around the sites of fracking locations to remove some of the toxins, specifically hydrocarbons found in produced water. Using this method, produced water can be inoculated with these bacteria to potentially reduce the number of toxins present in the water. In this experiment, bacteria from well sites in the Appalachian region were cultured and used to study their effectiveness in treating produced water. A pure culture of bacteria from Saskatchewan that has been shown to degrade hydrocarbons was also cultured. These samples were grown under three conditions; nutrient media with crude oil, nutrient media with produced water, and nutrient media with a mix of crude oil and produced water. A total of 36 samples were included, 3 replicates per treatment and bacteria culture. Water quality samples were taken and sent to the West Virginia University analytical lab where they were tested for strontium, barium and chloride ion concentrations in the samples taken at 0, 30 and 90 days. The results were then compared to the heat killed samples to determine if the bacteria had any effect on the produced water. It was found that both bacteria reduced the conductivity in samples with a mix of produced water and crude oil. It was found that bacteria from soil samples had a greater reduction of in ion concentrations in samples with just crude oil while the pure culture bacteria was more effective in samples with just produced water.

638710 Proposal of a Compostable Filter Medium Using Cellulose Fibers from Sugarcane Bagasse

Abstract:

The pandemic caused by the SARS-CoV-2 virus known as COVID-19 has generated an increase in the consumption of clothing focused on reducing the risk of contagion and respiratory diseases, with masks being the fundamental article to combat the spread of COVID-19 According to OMS, the problem associated with the exponential increase in the use of surgical masks is related to the 100% increase in demand in companies such as IWR Comercial, a similar case of 3M, being import companies in Colombia and in the case Colombian companies such as Industrias Saver, it is estimated that 90% of the masks can end their useful life in streets, landfills or even in the ocean, generating an environmental problem due to the material from which they are manufactured, which indicates more than 400 years to its disintegration.

In the first place, the proposal of a filter medium using cellulose fibers is developed from the agroindustrial residue of sugar cane, to be incorporated in cloth masks and contribute to filtering efficiency, being for Colombia the cultivation of sugar cane the second to the greatest extent, based on contamination caused by surgical masks and the uncertainty associated with the filtration of cloth masks, new alternatives that are biodegradable or suitable for compostable processes are created, using lignocellulosic compounds used in non-industrial environments. doctor. From this, alternatives have been developed such as the production of masks based on cassava starch or with banana peel, among others.

To obtain the filter medium, it begins with the extraction of cellulose from the alkaline hydrolysis at 5% of the cane bagasse fibers, generating the cellulose pulp, from which the percentages of lignin, cellulose, moisture, ash, real and apparent density and porosity using the Klason method, a methodology similar to the Kraft process, gravimetric methodologies.

After the characterization, a factorial system 2 is developed2 referring to additives, that is, glycerin at 20% and 50%, flour 7 grains 10% and 70%, with respect to the weight of the average filter 1.54 ± 0.19 g, being passed through the mixture of these variables in 20 ml of hot water and moistening the filter quickly to avoid tearing and / or rupture, improving flexibility and porosity when a glycerin concentration of 20% is obtained with 70% of flour 7 grains.

638343 Comparing Indoor Air Emissions from Cookstoves Using Plastic Derived Fuel Oil with Traditional Fuels

Abstract:

Combating the ongoing environmental crisis caused by mismanaged plastic waste is a global challenge. This problem is especially acute in developing regions due to a lack of recycling and waste management technologies. One way communities can combat this waste management challenge is by using the process of slow pyrolysis to convert the plastic waste into a liquid cooking fuel. The plastic waste can be collected by local entrepreneurs, converted into fuel oil using a small scale pyrolysis reactor, and sold for a profit. In addition to combating the mismanaged plastic waste, using this fuel in cookstoves can also help combat the public health issue caused by breathing in smoke from a cooking fire. Open fire cooking remains a common means of cooking in the developing world, and long term exposure to smoke can lead to chronic lung and eye health problems. The burden of these health problems fall disproportionately on women. Our hypothesis is that switching from wood fire cooking to using stoves fueled by liquid fuel produced from waste plastic will have a positive impact on indoor air pollution, including particulate matter, sulfur dioxide, and carbon monoxide. To test this hypothesis, a series of experiments was conducted. Particulate emissions (PM10 and PM 2.5) for the plastic derived fuel oil, traditional kerosene, and fire wood were measured using a Temtop M2000C Air Quality Monitor. Carbon monoxide was measured using a multigas detector. Cookstoves similar to those used in households in developing countries were used when conducting experiments. The results of these experiments indicated that Plastic Derived Fuel Oil is a cleaner alternative to kerosene and fire wood for household cooking. If implemented globally, the results of this research suggest that converting plastic waste into cooking fuel can not only reduce the amount of plastic waste entering the ecosystem, but can also combat the global public health problems caused by open fire cooking.

637656 Surfactant Effects in Irradiated, Hanging-Droplet, Aqueous-Phase Glyoxal/Ammonium Sulfate Aerosol Mimic System

Abstract:

The atmosphere contains aerosols, many of which are composed of organic materials. However, there is a level of uncertainty in regards to the aerosol in the atmospheric model due to its formation via secondary organic aerosol (SOA). SOA are formed via in situ interactions with volatile organic compounds (VOCs). Carbonyl-containing VOCs (CVOCs) such as Glyoxal (G) and MethylGlyoxal (MG) are SOA precursors. Products formed in dark chemistry of G and Ammonium Sulfate (AS) solution were seen not to be surface active resulting in no significant changes in the surface tension of the solution. G/AS solutions were aged for 24 hours suspended and exposed to ultraviolet light (UV). Through ANOVA analysis there was a statistical difference between irradiated and non irradiated samples to the 95% confidence due to p values being greater than .05. However the analysis on the concentrations showed that there were no statistical differences in the concentrations tested due to being in the same tukey group.Currently, it is being examined if the effects of exposure of UV lights to G/AS reaction systems change with the presence of Nonanoic Acid and Sodium Dodecyl Sulfate (SDS). In addition, the bulk concentrations will be examined to see if it affect any surface-dependent effects that occur due to the presence of extra surfactants

638652 Comparing Low- and Mid-Cost Particulate Matter Sensor Performance Indoors

Abstract:

Until recently, particulate matter (PM) monitoring instruments have been very expensive (>$10,000s), only permitting indoor air quality research over short periods of time in a select few locations and prohibiting the general public from assessing their indoor air quality. However, several low-cost ($100s) and mid-cost ($1,000s) sensors have recently entered the commercial market and are being widely deployed by the public. While low- and mid-cost sensors have expanded access to indoor air quality information, it is crucial that their limitations are fully understood so that they can be used to successfully mitigate exposure to poor air quality. In this study, we compare PM data reported by low-cost (Plantower PMS) and mid-cost (QuantAQ MODULAIR-PM) PM sensors in five different indoor environments on Georgia Tech's campus: two classrooms, two academic common spaces, and a cafeteria. PM data were collected in the five indoor environments during a portion of the Fall semester every minute for three PM size bins: PM_1, $PM_{2.5}$, and PM_{10}. Compared to the MODULAIR-PM, the Plantower PMS only measures approximately 45% of PM_1, 70% of $PM_{2.5-1}$; 72% of $PM_{10-2.5}$ across all five indoor environments. Furthermore, the low-cost sensor data is only strongly correlated for PM_1 ($R^2 > 0.93$) but not $PM_{2.5}$ ($R^2 < 0.13$) or PM_{10} ($R^2 < 0.19$) across all environments. Therefore, while the nephelometer performs well and can be used to assess exposure to indoor particles with diameters less than 1.0 micron, it fails to accurately assess exposure to indoor particles of larger diameters. This is very important because the mass fraction of large particles to all particles is significant (7% on average) in all five indoor environments evaluated in this study. The Plantower PMS is expected to miss larger particles because the sensor's capability is limited by its detection range: it only detects particles with diameters between 0.3 to 1.0 micron, a fraction of PM_1, and extrapolates this data to $PM_{2.5}$ and PM_{10}. Conversely, the MODULAIR-PM sensor can measure particles with diameters between 0.35 to 40.0 microns. With over 10,000 low- and mid-cost sensors installed globally and increasing scientific publications relying on low-cost sensor data, it is imperative for users to understand these sensors' limitations to draw solid conclusions about the measured air quality.

638698 Microplastic Occurrence in Water Matrices: Metadata Analysis and Occurrence Trends

Abstract:

Microplastics are polymer fragments < 5mm either manufactured in small sizes or resulting from material degradation and have been detected in water matrices from around the world. There is concern that ingested microplastics may pose a hazard to humans or wildlife. To inform mitigation strategies, research has been conducted to understand the prevalence of microplastics in aquatic environments. Standard methods for extraction of microplastics proposed by NOAA in 2015 were improved upon by the recent publication of ASTM methods, while standards for polymer identification are still in development. This has resulted in a wide variety of methodologies being reported with varying degrees of reliability and reproducibility. The aim of this study was to analyze current literature to understand (1) microplastic occurrence in various water matrices, (2) the effectiveness of different water treatment processes, and (3) the potential impact of study design on results. Metadata and results were collected from a total of 80 published studies on microplastic observations in drinking water, stormwater, surface water, and wastewater and incorporated into a database with concentration data for >2300 samples. Criteria for inclusion in the database required the use of chemical analysis (e.g., FTIR, PyrGC-MS, Raman) to confirm microplastic polymer type. Currently, the database includes 50 studies of microplastics in wastewater that were found to meet the criteria for inclusion in the database, compared to 19 for drinking water, and 11 for stormwater. A variety of study quality assurance and control practices were reported: 75% of studies reported using blanks (63.5% of which had some contamination) and 20.2% used matrix spikes to confirm particle recovery. With respect to extraction techniques, most studies applied wet peroxide oxidation (52.2%) to remove surface contaminants and reduce interferences from easily oxidizable particles, 39.2% applied some form of density separation, while few applied the ASTM recommended proteinase and lipase enzymatic degradation for complex matrices. Researchers applied analytical techniques to confirm a set number or percent of visually identified particles, particles stained with lipophilic dyes, or all particles. The potential impact of these extraction and analysis choices on results will be explored. Secondary wastewater treatment plants achieved an average of 88.8% removal of microplastics as opposed to tertiary wastewater treatment plants which produced an average removal of 92.2%. This 3.4% increase in the extent of treatment indicates that current treatment processes after secondary treatment may have a negligible effect on the removal of microplastics.

638006 Water Uptake of Monoterpene Aerosol Mixtures Produced in an Environmental Chamber

Abstract:

Understanding secondary organic aerosol (SOA) properties is essential in understanding cloud formation, which represents a significant uncertainty in the context of global climate modeling. Monoterpenes and sesquiterpenes are classes of semi-volatile organic compounds (SVOCs), of both biogenic and anthropogenic origins, emitted into the atmosphere. These compounds, when oxidized, contribute to the formation of SOA, resulting in the development of photochemical smog. This represents a significant burden, as it has been found that half of all total fine organic aerosol can be attributed to monoterpene oxidation (Zhang et. al, https://doi.org/10.1073/pnas.1717513115). The resultant aerosol particles may also act as cloud condensation nuclei and contribute to cloud growth. Understanding the water uptake ability, or hygroscopicity, of these compounds improves our knowledge of the radiative effects of smog, which can be used to improve climate models. This study measures the supersaturated droplet growth of fourteen SVOCs throughout controlled exposure to ozone in Bucknell's environmental chamber through the use of cloud condensation nuclei counting (CCNC). Data from CCNC was collected to produce a single parameter hygroscopicity, or kappa, value for each compound (Petters et. al, https://doi.org/10.5194/acp-7-1961-2007). Given that these components rarely occur in the atmosphere in isolation, an understanding of their group hygroscopicity will prove more valuable than any individual measure. To model this, mixtures of higher quantity components were run, namely 2 and 5 SVOC-component experiments. Included compounds were informed by variation in kappa value observed in individual runs, so as to provide a span of mixture types, from similar (high/high) to contrasting (high/low) in value. While not every possible combination was chosen for experimentation, mixtures were intentionally selected so as to provide a wide base of meaningful data. Preliminary results show that as increasingly many SVOCs are introduced into the reaction mixture, kappa values trend towards a central value, approximately 0.14. This proves promising for further work on increasingly higher component mixtures.

637674 Modeling Salt Concentration Effects on Glyoxal/Ammonium Sulfate Dark Chemistry Kinetics

Abstract:

Aerosols have a significant impact on the unsteady-state energy balance of the atmosphere, contributing both warming and cooling effects to global radiative forcing. Light-absorbing brown carbon compounds are formed from reactions of carbonyl-containing volatile organic compounds (CVOCs) such as glyoxal and methylglyoxal which are present in the atmosphere. This research aims to determine if there is a salt concentration dependence on modeled rate constants for the glyoxal/ammonium sulfate dark chemistry reaction system. Experiments are carried out to simulate the glyoxal/ammonium sulfate dark chemistry reaction at different salt (ammonium sulfate) concentrations. The data was compared to the Triple Gaussian model developed in previous studies and rate constants for the system were inferred assuming first order kinetics. Statistical analysis was carried out to determine if the rate constants scale with respect to salt concentrations. The information obtained could provide insight into organic aerosol behavior by furthering our understanding of the influence of salt concentration on rate constants for reactions occurring in the atmosphere.

638424 Designing a Portable Particulate Matter Monitor

Abstract:

Air pollution affects billions of people around the world, yet policy makers often have difficulty obtaining precise, local air pollution measurements. Local air pollutant concentrations can vary dramatically depending on the possible presence of sources (factories, busy roads, residential wood burning). Though fixed-site air pollution monitoring is available, this method does not give a reliable measure of how $PM_{2.5}$ (particulate matter under 2.5 micrometers) concentrations vary with respect to space. We developed a portable method to measure local air pollution with an Alphasense OPC-N2 particulate matter monitor while commuting by foot, bicycle, or car. We collect GPS coordinates via the mobile app, Strava. Then we use RETIGO, EPA's Real-Time Geospatial Data Viewer, to analyze the data and generate maps of the spatial variability of $PM_{2.5}$ concentrations along the sampling route. Preliminary data collected along a busy road yielded higher concentrations of $PM_{2.5}$ compared to areas with less traffic, confirming that traffic is contributing to local air pollution. This approach provides an intuitive visualization of possible local pollution sources that is accessible to both researchers and the general public. When validated by more experiments, this tool can be an easy way to measure local air pollution. This project can have important implications for how local air pollution can be measured and managed, which can greatly impact public health and environmental equity on a larger scale.

637916 Evaluation of TROPOMI Surface UV Irradiance in the Continental United States : Preliminary Results

Abstract:

Solar ultraviolet (UV) radiation at short wavelengths ranging from 280-400 nm is well known to cause harmful effects in humans. Sunburn, cataracts, and increased risk of skin cancers including melanoma are all possible effects of exposure to UV radiation. A steady increase in the amount of people diagnosed with serious skin cancers such as melanoma has been observed worldwide, with rates almost doubling since 1999. In order to understand and study the health effects of UV radiation, information on the geographical distribution of radiation reaching Earth's surface is needed. As the successor of Ozone Monitoring Instrument (OMI), which has been used since 2004 to monitor surface UV radiation levels, the TROPOspheric Monitoring Instrument (TROPOMI) satellite, launched in 2017, will continue the efforts to monitor the surface UV radiation levels. Although TROPOMI is able to provide global measurement of UV levels, measurements can only be taken once per day and complications with cloud cover and aerosol levels can impact the accuracy of the measurement. Some efforts have been made to verify TROPOMI data in Europe and Australia; however, no organized effort has been taken to make this verification for the United States. In this work, we will use the ground-based sensor network through the Colorado State University Natural Resource Ecology Laboratory (NREL) to evaluate the TROPOMI surface UV radiation. This network records UV levels at 30 ground sites across the US every 15-20 seconds. This project looks to use Python to conduct the data analysis. As part of this effort, TROPOMI data was downloaded and regridded to more standard grids, which would also be convenient for the community to use. Ground data from the NREL was downloaded, and quality control was performed to refine the data. Time series plots were created for many of the ground sites. Maps were then created to visualize the levels of UV radiation across the US. Finally, we processed the TROPOMI data and spatially and temporally matched it to the corresponding ground sensor data to evaluate the TROPOMI data performances. Further analysis of the trends of the satellite and ground-based surface UV data is planned for the future.

638049 On the Development and Setup of a Hygroscopic Tandem Differential Mobility Analyzer for Aerosol Studies

Abstract:

Most particles in the atmosphere take up water when exposed to increasing relative humidity, and this behavior is described and quantized by hygroscopicity parameter. Hygroscopic tandem differential mobility analyzer (HTDMA) can be used to measure the physicochemical properties of aerosol particles and study phenomena that lead to size changes in submicron aerosol particles. (Rader and McMurry, https://doi.org/10.1016/0021-8502(86)90031-5). Evaporation, condensation and chemical reactions within airborne droplets are examples of such phenomena. Specifically, it measures how aerosol particles of different initial dry sizes grow or shrink when exposed to changing relative humidity (RH). HTDMA uses two different mobility analyzers (DMA) and a humidification system to make these measurements. One DMA selects a narrow size range of dry aerosol particles, which are exposed to varying RH conditions in the humidification system. The second humidified DMA scans the particle size distribution after the particles have been conditioned by the humidification system. Scanning a wide range of particle sizes enables the second DMA to measure changes in size or growth factor (growth factor = humidified size/dry size) due to water uptake by the particles. A condensation particle counter (CPC) downstream of the second DMA counts particles as a function of selected size in order to obtain the number size distribution of particles exposed to different RH conditions. To build up a HTDMA design, two DMA was connected, with one humidifier (along with one RH meter) in the second DMA sheath flow, and one humidifier (along with one RH meter) in the particle flow line before aerosol entering the first DMA. The outflow from the second DMA was running in to a CPC, and the data collected from CPC was inputting into software for further analysis on particle size and distribution. Different size of humidifiers (mainly MH-series Nafion tubes) was tested under different flow rates (0.3 lpm, 1.5 lpm, 3 lpm, 15 lpm). The two humidifiers provide the close %RH in particle flow line (0.3 lpm or 1.5 lpm) and in the sheath flow line (3 lpm or 15 lpm) were used in the design. The HTDMA design was tested running through pure ammonium New HTDMA software was also developed to analyze the data on hygroscopicity. Instead of using a manually selected mode, the new software will autodetect peaks and it fits a log-normal distribution producing a more reliable average particle size.

638235 Optimization of Parameters for the Analysis of Enzyme-Induced Carbonate Precipitation (EICP) Cemented Sands

Abstract:

Erosion in sloping sands has been known to be seen on the side of highways, bridges, cliffs and more. One of the majors causes of erosion is storm water runoff. In order to try and improve the erosion caused by storm water runoff seen in sloping sands is by cementing the sand using Enzyme-Induced Carbonate Precipitation (EICP). When the EICP solution is used to improve erosion, it will have precipitated calcium carbonate into the sand. The amount of calcium carbonate is important as it is what binds individual sand grains together causing cementation. Every type of sand will be different causing the EICP solution to precipitate different amounts of calcium carbonate depending on the conditions of the sand. The calcium carbonate content that is found in the sand after cementation can be determined by acid digestion. The optimal optimal parameters for the acid digestion procedure have not yet been determined. For this study, different parameters that make up the acid digestion procedure were modified in order to study the variation of calcium carbonate content in the EICP cemented sand. The goal of this study is to develop a standard protocol for the determination of calcium carbonate content using acid digestion.

638356 Phytoremediation of Radionuclide-Contaminated Soils

Abstract:

Northwestern New Mexico has many abandoned uranium mine sites, some with high levels of radionuclide contamination, such as uranium and radium. Contamination is generally highest around former ore mills and mine tailings, with addition spread from water and wind erosion of the soils.

To remediate the soil of one such mine site with moderate levels of contamination, a series of greenhouse tests were conducted using soil from the site. Three growth studies were completed: two to determine which plants grew the best in the soil and which growth inputs were needed (fertilizer, irrigation, etc.) and one to perform a radionuclide mass balance. To facilitate proper drainage, the plants were placed in a 1:1 volume mixture of inert sand and soil. The three selected plant varieties were guar (*Cyamopsis tetragonoloba* var. Kinman), sunflower (*Helianthus annuus* var. Cobalt 2), and hemp (*Cannabis sativa* var. Samurai.) Seeds were planted directly into the soil and allowed to grow until the plants reached maturity and both flowers and seeds presented. Aboveground biomass and seeds were harvested for analysis. Plants were washed using deionized water to remove residual soil. Plant biomass was combusted in a muffle furnace and the ash was acid-digested for total metals analysis. The seeds were processed in an oil press to separate out an oil fraction that was also acid digested. Metals, including uranium and radium, were quantified using a combination of inductively-coupled plasma optical emission spectroscopy (ICP-OES), inductively-coupled plasma mass spectrometry (ICP-MS), or multi-collector inductively-coupled plasma mass spectrometry (MC-ICP-MS.)

Mass balances from the greenhouse study were used to estimate the time required for the concentration of radionuclides to be reduced to below levels required by the U.S. EPA. Results indicate that the three plant species do uptake radionuclides and fractionate the metals into the non-oil components of the plant. Soil remediation timescales are on the order of centuries; however, other benefits of phytoremediation, such as soil cover and value-added use of the biomass for renewable energy may justify continued field development.

637057 The Impact of Lignin Content on the Pyrolysis Time-Temperature Curves during Biochar Production from Household Biowaste

Abstract:

Biochar can be used as a carbon sequestration method or for other value-added applications in industries like agriculture or environmental control. Organic wastes can be converted to biochar by heating in an inert environment in a process known as pyrolysis. This process can be carbon negative as in pyrogenic carbon capture and storage (PyCCS). We studied the influence of feedstock lignin content on the pyrolysis process using ten different feedstocks, including walnut shells, banana peels, and sunflower seed shells. Pyrolysis was conducted at 500°C in a tube furnace with two thermocouple probes monitoring the temperature of the upper and lower regions of the tube. We have found that different feedstocks have different trends in their pyrolysis time-temperature curves, and there can be both exothermic and endothermic regions in the curves. The data indicates that the maximum change in temperature decreases linearly as the lignin content increases, and reaction time increases as lignin percentage increases. In an additional study, pyrolysis was conducted at various temperatures on three different feedstocks: walnut shells, cardboard, and sawdust. We have found expected trends within feedstock types, such as the solid yield increasing as the pyrolysis temperature decreases. We will present the endothermicity of pyrolysis as both a function of feedstock and temperature.

638490 Application of Machine Learning to Identify Key Parameters in Chemotactic Bacterial Distributions Near Oil Ganglia in Porous Media

Abstract:

Nonaqueous phase liquids (NAPLs) are pollutants that can persist underground for decades due to their low water solubility. Bioremediation may provide a cheap and effective solution for removing NAPL contaminants underground. Chemotaxis, a mechanism bacteria use to detect and move towards higher chemical concentrations, including NAPLs, may enhance the overall rate of biodegradation. Since bacteria consume NAPLs as food, greater accumulation of chemotactic bacteria near the NAPLs suggests higher efficacy of bioremediation. Our study applies a Machine Learning algorithm to evaluate factors that may influence bacterial transport and chemotaxis in a dual-permeability microfluidic device in the presence of trapped NAPL ganglia. From our experimental observations, we hypothesize that bacterial accumulation is impacted by the dimensions of the micro-pocket, flow rate, time elapsed since the release of bacteria, and locations of NAPL ganglia. Machine Learning analysis suggest the length of the micro-pocket is an important feature for regulating bacterial accumulation. Longer micro-pockets make NAPL gradients more robust because fluid velocity slows down near the NAPL ganglia. Therefore, bacteria may accumulate at higher rates upon sensing these gradients. Linear Regression is a practical starting point to identify trends in numerical variables. In future studies, we plan to test more advanced Machine Learning algorithms, such as neural networks. At the same time, we will obtain more experimental data from microfluidic devices to improve our models. Machine Learning is expected to be a useful tool to understand the transport of chemotactic bacteria in oil-contaminated porous media and improve applications of chemotaxis in bioremediation.

638441 Biochar Production from Cardboard Feedstock

Abstract:

Biochar can be used to sequester carbon, as a low-carbon fuel, or for applications such as regenerable agriculture. Biochar research commonly focuses on agricultural and forestry waste products, municipal waste, and food waste, but research for cardboard as a feedstock is limited. Although cardboard can be recycled, the process is limited by diminishing material quality of 10% while other materials, such as paper or aluminum, have negligible material loss. Our research investigates a new feedstock, corrugated cardboard, as a possible biochar feedstock. Biochar from corrugated cardboard was produced in a pyrolysis furnace reactor as well as in a Thermogravimetric Analyzer (TGA). Our analysis of mass as a function of temperature determined that the pyrolysis process for our feedstock sample begins at 350°C. Pyrolysis was performed in an electronic furnace at temperatures between 350°C and 500C°C leading to char yields between 26% and 19%. At 500°C, char yield from the pyrolysis reactor was 19.6%, similar to char yields of 21.6% from the TGA. Of nine previously studied bio-waste feedstocks for char production, (walnut shells, pistachio shells, coconut shells, corncob, pumpkin seeds, sunflower seeds, sawdust, and coffee grounds), the cardboard yields were among the lowest when measured at a pyrolysis temperature of 500°C. The ash, or inorganic content, of the cardboard char was found to be 11.7% of the biochar, higher than that of previously studied feedstocks like nutshells (3.8-6.0 wt% ash in biochar) and lower than feedstocks like corncob and sawdust (23.3-26.5 wt% ash in biochar). This research shows that cardboard can be a potential feedstock for biochar and pyrolysis can be used as a waste management tool to produce value from previously recycled cardboard.

638450 Acid Tolerant CAH Bioremediation: Combined Experimentation and Modeling of Hydrogel Encapsulated Biobeads

Abstract:

Trichloroethylene (TCE) and other chlorinated aliphatic hydrocarbons (CAHs) are a widespread issue in groundwater due to their acute toxicity, carcinogenicity, and volatility. Long term TCE exposure at low levels has been linked to kidney cancer, non-Hodgkin's lymphoma, cardiac defects, and death. Anaerobic reductive dechlorination by microorganisms (bioremediation) is a cost-effective solution to treat TCE; however, during remediation acid (hydronium ions) can accumulate to levels that inhibit or are lethal to degrading microorganisms and halt degradation in a more hazardous state.

Biodegradable, spherical hydrogel encapsulated microorganisms, also known as "biobeads", allow for tuned control of mass transport and the ability to couple non-biological processes to reduce the acid generated during bioremediation. Our research group has concentrated on characterizing mass transport behavior of CAHs and hydronium ions in different polymer hydrogel systems, creating methods to characterize alteration of microorganism viability during encapsulation, and developing computer simulations of coupled reaction-diffusion processes to optimize biobead design.

Mass transport of CAHs, including TCE, cis-1,2-dichloroethylene (cDCE), and vinyl chloride (VC) in hydrogels were measured spectrophotometrically from temporal measurements using a custom diaphragm cell. Computer simulations were developed in Matlab from a central finite difference method using published kinetic parameters for CAH degradation and pH inhibition. Measured diffusion rates of CAHs were used in the model. A sensitivity analysis was performed using cell concentration, CAH diffusivity, and biobead radius.

Effective diffusivity of TCE, cDCE, and VC was enhanced in PVA/Alg membranes over strictly aqueous diffusivity by -10±25%, +39±10%, and +69±17%, respectively. The difference in diffusivities between CAHs may be explained by dipole interactions in correlation to the symmetry of the molecules. Measured diffusivity rates were used in the reaction/diffusion model. Even at fully saturated external TCE concentrations, simulated microbes degraded contaminants while encapsulated within the hydrogel. For a single biobead, without accounting for microbial pH inhibition, as cell concentration increased, generation and efflux of VC to the environment increased to a maximum then decreased and was dependent on hydrogel radius.

Measured diffusivities of CAHs represent the first characterization of mass transfer coefficients for CAHs in a hydrogel matrix and these parameters along with the reaction/diffusion model represent a critical component for engineering and optimizing a encapsulated bioremediation solution. Because the geometry of biobeads allow for simultaneous degradation of TCE, cDCE and VC and can be modified to respond to stimuli and environmental conditions, these hybrid materials represent a promising tool for remediation of CAHs.

638603 Using Plastic Derived Fuel Oil from Polyolefins As a Sulfur-Free Cooking Fuel

Abstract:

Polyolefin plastics are extremely detrimental to the planet because of the time it takes to decompose. Currently, we accrue about 300 million tonnes of plastic waste every year. Another worldly issue is the detrimental burden that women in third-world countries have from cooking over an open fire. From the results of GC/MS, plastic derived fuel oil has proven to be a cleaner cooking fuel than traditional kerosene because it is sulfur-free. Utilizing these results, both major issues could simultaneously be improved and/or solved. Burning firewood releases harmful, climate-warming emissions that causes forest degradation, which in return, causes losses in erosion control, biodiversity, and flood protection. Plastic derived fuel oil could potentially be a cleaner alternative that has more environmental-friendly emissions and while also helping women who carry these health-related burdens that come from using firewood. The principles of appropriate technology can be applied to allow developing world communities to produce plastic derived fuel oil themselves. Results will be presented consisting of a GC/MS analysis of fuel oil produced from polyolefin plastic waste and compared with traditional kerosene cooking fuel. The results will show that cooking fuel from plastic waste is a viable alternative to biomass and traditional kerosene.

637726 Effects of Pyrolysis Temperatures on CO2 Adsorption Rates and Capacities of Biochar As an Alternative Solid Adsorbent for Carbon Capture

Abstract:

Due to rising atmospheric carbon dioxide (CO_2) levels, there is growing interest in removing CO_2 from ambient air through Direct Air Capture (DAC). DAC technologies typically capture harmful atmospheric CO_2 using liquid amines or solid adsorbents. Solid adsorbents for DAC can be costly to produce and maintain and can be environmentally harmful at the end of their lifetimes. Biochar is an inexpensive form of activated carbon made from biowaste with the ability to adsorb CO_2; therefore, it is a promising alternative solid adsorbent for DAC. Properties of biochars have been well-studied but not optimized for DAC. We investigated the CO_2 adsorption capacity of biochars between 25 and 50 degrees Celsius. We tested biochars made from various biowaste, including coconut shells, corncobs, sunflower seeds, walnut shells, pistachio shells, coffee grinds, and pumpkin seed shells. We tested the effects of pyrolysis temperature, CO_2 concentration during pyrolysis, moisture capacity, ash content, and lignin content on CO_2 adsorption rates and capacities. Walnut shells and sawdust biochars had a strong positive correlation between pyrolysis temperature and CO_2 adsorption capacity. As pyrolysis temperatures were increased from 500-800 degrees Celsius, walnut shell biochars demonstrated over a twofold increase in CO_2 adsorption capacity from 0.390 mmol/g to 0.815 mmol/g. Sawdust biochar pyrolyzed under pure N_2 at 800 degrees Celsius achieved the highest overall CO_2 adsorption capacity at 50 degrees Celsius (0.840 mmol/g). This value is highly competitive with current solid adsorbents used in industry for DAC. Our results indicate that CO_2 adsorption capacity might be independent of feedstock type when pyrolyzed at high temperatures (e.g. 800 degrees Celsius).

637878 Evaluation of Different Microbial Seeds on the Carbon Dioxide Production and Weight Loss of Sugarcane Bagasse Composting

Abstract:

Adsorption is a common tertiary wastewater treatment for industry in which natural, lignocellulosic, adsorbents have been researched regarding their potential for activated carbon (AC) replacement. Generally, these adsorbents have a lower adsorption potential, but are a more abundant and cheaper alternative to AC. Sugarcane bagasse is a promising material as a natural adsorbent. Once any adsorbent has reached its maximum capacity for industrial contaminants, it must be managed as a hazardous material. Currently, there is a lack of data on how these novel adsorbents will be either treated or managed after they are spent. Currently –composting is a viable disposal technology for lignocellulosic biomass and a known hazardous waste treatment. This study aims to initiate a series of experiments on compost-based biodegradation as a management and treatment method for spent, natural adsorbents using sugarcane bagasse as a model material.

The objective of this study is to evaluate several microbial inoculum sources; compost, activated sludge, cow manure and white rot fungi, for lignocellulosic biodegradation by monitoring carbon dioxide production and weight loss of a sugarcane bagasse adsorbent. Using forecasting to evaluate mass loss at 60 days, cow manure was forecasted as the best inoculum with a predicted 120% mass loss and white rot fungus consortium next with a 74% mass loss.

638642 Control of Gene Expression in *E. coli* with the Twister Sister Ribozyme

Abstract:

In the current study, the use of the recently discovered *twister sister* ribozyme was analyzed to determine its ability to control gene expression in *Escherichia coli*. Primer extension PCR was used to create a hammerhead ribozyme that could be introduced through the P1 stem. An anti-ribosomal binding site (RBS) sequence was also included in the ribozyme construct, and the ribozyme was inserted upstream of the *gfp* gene cloned into the pET28a(+) plasmid. Additional ribozyme constructs inserted through the P3, P4, and P5 ribozyme stems were also constructed. Plasmids were then transformed into *E. coli* BL21(DE3), along with controls of pET28a(+) only and pET28a-gfp without the ribozyme, and cells were grown at 37°C for 16 h. While the pET28a-gfp control produced a high level of fluorescence, all ribozyme constructs reduced fluorescence of the gfp to near background levels, indicating a reduction in gene expression. Because the twister-sister ribozyme is Mg2+-dependent, we also tested the effect of varying levels of MgSO4 in the growth medium. *In vitro* transcription and activity assays demonstrated that the twister sister ribozyme constructs created in this study do possess catalytic activity, indicating that they may be of use in controlling gene expression *in vivo*.

638488 Prototyping of Genetic Parts through Clostridia-Based Cell-Free Platforms

Abstract:

Climate change is a multi-faceted and growing global crisis that is exacerbated by greenhouse gas emissions such as CO2 and CO. A large source of such pollution is the release of waste gas streams from industrial scale productions such as coal-powered plants and steel manufacturing. Such emissions negatively impact the environment and methods to reduce pollution must be leveraged across industry. Biotechnology and strain engineering have been rising sectors in recent decades as microbes hold great potential for undergoing complex chemical reactions, however, engineering organisms is an extremely lengthy and costly process from the growth of cell cultures to arduous genetic modifications. These slow time frames result in long periods of research for new technological strategies as well as for any large-scale implementation. The utilization of cell-free systems and cell-free protein synthesis (CFPS) can greatly accelerate research timelines by producing proteins and mimicking metabolic pathways *in vitro.* This allows for rapid experimentation and optimization of genetic parts, mutations, and enzymatic pathways prior to *in vivo* strain growth, with potential for high throughput screening strategies as well. I applied this framework of cell-free prototyping towards Clostridia strains: anaerobic bacteria with unique metabolic pathways that are capable of converting CO2 to short chain organics compounds. As this bacterium has just recently begun to enter commercial plants, its cell-free platform and distinctive pathways are much less studied in comparison to other microbes such as E. coli. I revealed ranges of CFPS efficiency and demonstrated genetic protein product control through successful modification and replacement of key genetic parts such as promoters, terminators, and ribosomal binding sites. This work provides insights and expands our knowledge into the genetic control of Clostridia cell-free systems, multi-step synthesis development, and progresses towards sustainably circular environments.

637647 Identification of Biologically Active Compounds That Drive DACH1 Expression

Abstract:

Diminished expression of the transcription factor Dachshund homolog 1 (DACH1) was recently identified as a major kidney injury susceptibility factor for both podocytes and tubular cells. Therapeutic strategies that augment DACH1 expression in these cell types are high yield approaches that may prevent chronic kidney disease (CKD) progression. Transcription factor modeling of the DACH1 promoter predicts the presence of several glucocorticoid response elements. We show that glucocorticoid (GC) administration caused early and sustained induction of DACH1 expression in both human podocytes and human embryonic kidney (293T) cells. While we speculate that DACH1 may be a central transcriptomic regulator of the beneficial GC response, GCs are riddled with systemic toxicities that limit their use. With this in mind, we designed a high throughput screening assay to identify novel biologically active compounds that drive DACH1 promoter activity. To do this, we cloned the DACH1 promoter, including the 3kb immediately upstream of its ATG start site, into a luciferase reporter construct (DACH1-Luc). We are currently developing stably transfected 293T cell lines that carry the DACH1-Luc or control vector alone. Our plan is to use these cells to screen a 4,000 compound library from the National Cancer Institute (NCI). Compounds will be selected based on their ability to drive luciferase activity, after which their biological activity will be confirmed by western blotting for DACH1 protein in compound-treated cells. Our goal is to test the therapeutic potential of select lead compounds in mouse models of CKD and ultimately in human clinical trials.

636949 Identifying Escape Mutants on the Sars-Cov-2 Spike Receptor Binding Domain Using Yeast Screening

Abstract:

The increased prevalence of SARS-CoV-2 variants like Delta (B.1.617) demonstrates the need for understanding the impact of novel mutants on the human antibody response, with impacts both for development of vaccine boosters and effective monoclonal antibody therapies. The SARS-CoV-2 Spike (S) receptor binding domain (RBD) is primarily responsible for infecting human cells by binding to human angiotensin-converting enzyme 2 (ACE2). Blocking this binding is a main correlate of protection, and so vaccines are developed primarily by the ability to elicit high amounts of antibodies which block this binding, thus neutralizing infectivity of the virus. However, current circulating variants have several RBD mutations which minimize neutralization by many of these antibodies. This work presents a complete experimental pipeline for identifying potential antibody escape mutations. We tested multiple antibodies, including CC6.29, CC12.1 and CC12.3, with yeast screening and deep mutational scanning. A competitive binding experiment was used to test the antibody's ability to bind in the presence of ACE2 after a mutation is introduced on the S RBD. Following the experimental procedures, custom software was developed to read and filter through output data from deep sequencing. The software also performs a statistical analysis to identify potential escape mutants. A Microsoft Excel per-position heatmap is generated, highlighting each position in the library that was identified as an escape mutation. With this approach, certain escape mutant positions repeatedly appeared for multiple antibodies: K417, D420, Y421 and Q493. Many of the identified positions were located on the perimeter of the ACE2 binding site on the S RBD. A primary limitation of these experiments is that the antibody tested must bind competitively with ACE2 on the S RBD. If the antibody binds to another region of the S RBD, then this method for identifying escape mutations will be insufficient. Even so, many antibodies currently being studied directly compete with ACE2 on the S RBD and these methods provide a relatively quick and reliable approach for determining escape mutations. Having the ability to test for escape mutants given an antibody will be imperative for better understanding this virus and to the future development of monoclonal antibody treatments for SARS-CoV-2.

637431 Epigenetic Effects of Extracellular Matrix Stiffness on Cardiac Fibrosis

Abstract:

Aortic valve stenosis (AVS) is a chronic, deadly disease that occurs in at least two percent of patients over 60 years of age in western countries. In AVS, the aortic valve leaflets thicken and stiffen as they become calcified. This results in disrupted blood flow as the valve can no longer open and close normally; over time, the chronic stiffening of the valve leaflet leads to enlargement of the left ventricle and ultimately contributes to heart failure. From a cellular perspective, AVS occurs as quiescent valvular interstitial cells (VICS) activate to persistent myofibroblasts. These myofibroblasts are characterized by an increase in pro-calcific factors and especially in alpha smooth muscle actin (αSMA). Mechanical and secretory changes that accompany this phenotypic change have been documented, but the epigenetic regulation of this transition has not been thoroughly investigated. Chromatin remodeling is an epigenetic change where histones can regulate the accessibility of chromatin. Additionally, chromatin remodeling has been strongly linked to cell fate determination in VICs, indicating that epigenetic mechanisms are involved. The *in vitro* culture stiffness has also been shown to play an extensive role in the activation of quiescent VICs to myofibroblasts. Hydrogels pose a unique opportunity to study this phenotypic transition through mimicking diseased tissue versus healthy tissue extracellular matrix (ECM). These covalently crosslinked polymer networks can be tailored to specific stiffnesses, and networks can be controllably softened to replicate these tissues. By utilizing photo-degradable hydrogels to mimic a stiff, fibrotic ECM versus a healthy, soft ECM, the epigenetic changes regulating AVS can be investigated. Assay for Transposase Accessible Chromatin sequencing (ATAC-seq) was used to investigate changes in the accessibility of chromatin in combination with RNA sequencing (RNA-seq) to assess the downstream effects of changes in chromatin accessibility for quiescent VICs versus myofibroblasts. Differentially expressed chromatin regions were then compared to RNA-seq results from stiff and soft hydrogel VIC cultures to identify possible regulatory genes. Four genes were found to be differentially expressed that directly relate to changes in chromatin accessibility: ACTA2, SERPINE1, ID3, and PAMR1. Additionally, 18 transcription factors (TFs) were implicated in this cellular transition. We hypothesize that these genes may be involved in myofibroblast regulation, and therefore regulating these will lead to significant changes in myofibroblast activation when cultured on the appropriate stiffness. These epigenetic changes provide new insight into potential mechanisms that regulate the stiffness-induced transition from quiescent VICs to myofibroblasts and that therefore may regulate AVS.

638037 A Mechanistic Study of Lysosome Infiltration of N-Dihydrogalactose Chitosan to Activate the Stimulator of Interferon Genes (STING)

Abstract:

In the humans, viral respiratory tract infections are the most common types of infection. Successful viral vaccination requires the development of antibodies (humoral immunity) and T cell responses (cellular immunity). Adjuvants are added to vaccines to sustain the antigen and enhance the cellular and/or humoral immunological responses to the targeted pathogen. However, current adjuvants approved for human use are not ideal for viral pathogens such as influenza and SARS viruses as they only initiate humoral and not cellular immune responses. Type I interferons (IFNs) are key to antiviral immunity and protective cellular responses. We have developed a novel biosynthetic polymer, N-dihydrogalactochitosan (GC), that drives the production of type I IFN and IL-1β responses in dendritic cells (DCs). DCs are professional antigen presenting cells (APCs) that direct cellular immune responses. To determine if GC works well as a stimulant/delivery method for vaccination against viral pathogens, we combined GC with the recombinant nucleocapsid and spike proteins from SARS-CoV-2 and vaccinated human ACE2 transgenic mice two times. Preliminary data revealed that this vaccination is highly protective against COVID-19 viral challenge, prompting us to uncover the mechanism by which GC activates DCs. In DCs, the stimulator of interferon genes (STING) is required for GC to induce type I IFNs and an inflammatory form of cell death (ICD) that activates caspase-1 to cleave IL-1β and gasdermin-D. Using live cell imaging we discovered that many DCs that phagocytose GC swell and rupture, and this is dependent on the lysosomes upstream of STING activation. We hypothesize that the lysosome digests GC into smaller fragments which them leak into the cytosol where it interacts with STING to trigger the production of type I IFNs. We further hypothesize that the strong binding of GC to STING results in STING trafficking to the lysosome to trigger lysosomal cell death, which recruits and enhances the inflammatory response of neighboring DCs. To examine whether GC is digested into smaller fragments by lysosomal enzymes, lysosomes were isolated from DC2-4 cells. Lysosomal extracts were then incubated with GC at varying concentrations and then analyzed via an SDS-Page gel to examine fragmentation of GC. To examine STING trafficking to the lysosomes, subcellular fractionation on GC stimulated DC2-4 cells was performed. Western blots were used to examine STING localization in the cytoplasm, lysosome, membrane bound organelles, and the nucleus. Our data will provide insight into the mechanism in which GC activates DCs to drive cellular immunity.

637770 Collagen mRNA Perturbations in Embryonic Chick Heart By Single Cell Sequencing

Abstract:

Introduction: As the most abundant protein within the animal kingdom, collagen's role in the development of embryonic hearts is of the utmost importance. Previous studies have shown as embryonic heart develop, collagen levels increase with stiffness. This stiffness combats the mechanical stress due to the heart beating. If heart tissue damage occurs, the cells are unable to regenerate, collagen forms scar tissue and could lead to heart failure. It is likely that collagen perturbations after DNA damage occurs. In treating embryonic chicken hearts with blebbistatin (blebb), a myosin II inhibitor, the heart stops beating and there is an overall decrease in mechanical stress. This decrease in mechanical stress corresponds with lower collagen levels. This study aims to use single cell sequencing (scRNA-seq) and differential analysis of collagen genes across cell types and drug treatment groups to gain insight about collagen response to various stimuli.

Materials and Methods: Embryonic chick (Gallus *gallus*) hearts were extracted after 5 days of development. The chick hearts were assigned into 4 drug treatment groups: Dimethyl sulfoxide (DMSO), blebb, blebb with blebb washout(blebb+wo), and H_2O_2. DMSO and H_2O_2 serving as negative and positive control for DNA damage respectively. scRNA-seq data was analyzed using the program Seurat. All data was normalized and underwent Uniform Manifold Approximation and Projection (UMAP) to identify cell clusters. These clusters were identified using biomarkers found in literature and previous studies. Differential analysis was conducted with the default Wilcox test. Second Harmonic Generated (SHG) images visualized collagen and myosin fibers within a tissue.

Results and Discussion: Using Wilcox test differential analysis, significant changes in collagen expression were determined. All treatment groups showed significant perturbations of collagen. Compared against DMSO, Blebb+wo had a significant decrease ($p<2.5e-6$) in expression of 2 collagen proteins (Figure 1B-i). However, H_2O_2 treated groups had both upregulation and down regulation ($p<.0002$) of 2 collagen proteins (Figure 1B-ii). After clustering the scRNA-seq data with UMAP, we identified epithelial, epicardial, cardiomyocytes, macrophages, erythrocytes, and valve cells (Figure 1A). Most collagen expression was found exclusively in the valve and epicardial cells. Cardiomyocytes expressed high levels of myosin; this is further confirmed with SHG. SHG indicated a stronger forward signal, corresponding with larger amounts of myosin compared to collagen.

Conclusions: This study was successful in the identification of cell types and observing the perturbations of collagen. Further analysis should find biomarkers for DNA damage that can be seen using single cell sequencing.

637768 Isolation of Rodent Microglia to Assess Gene Silencing and Drug Targeting

Abstract:

Aging is the strongest risk factor for neurodegeneration, with a majority of the elderly population experiencing cognitive decline. While many factors drive disease progression, impaired microglial dysfunction has recently emerged as a common hallmark of many age-associated neurodegenerative disorders. As the resident immune cell of the central nervous system (CNS), microglia serve important functions including monitoring the CNS for pathogens and apoptotic signals. However, in disease states, chronic overactivation of microglia contributes to neurotoxicity and promotes disease progression, particularly in the context of Alzheimer's Disease. The potential to explore microglia immunomodulation in culture is an invaluable tool to help mitigate age-associated neurodegenerative disease. We established a protocol to isolate microglia from the rodent CNS and maintain them in culture. This efficient method involves dissociating mouse brains into single cells with high viability, followed by magnetic sorting using beads coated with the antibody cd11b, a microglia-specific marker. Microglial identity was confirmed by flow cytometry for cd11b and immunostaining for iba1. Microglia were maintained in culture for up to a week and retained high viability and cellular identity. Furthermore, the magnetically sorted microglia were highly abundant with over 6,000 cells and enriched in microglia-specific markers compared to the remainder of the dissociated cell population. To demonstrate the versatility of the protocol, microglia were similarly isolated from rat brains and successfully cultured. For future work, observing phenotypic changes in microglia after siRNA-mediated gene silencing and applying therapeutics to diseased murine microglia will be explored in hopes of improving our understanding of treatments for neurodegenerative disease.

638046 Non-Viral Delivery of CRISPR-Cas 9 in Neuro-2a Cells for the Stable Production of Natural Products

Abstract:

Neurological cells tend to be more resistant to gene editing techniques, but if gene editing is possible in neurological cells it could increase therapeutic solutions for neurological diseases such as Alzheimer's disease, Parkinson's disease or brain tumors. Specifically for Alzheimer's disease, the introduction of curcumin to the cell decreases the effects of the disease. Non-viral delivery of a plasmid encoding for the enzymes tyrosine ammonia lyase (TAL), 4-coumarate-CoA-ligase (4CL1) and curcuminoid synthase (CUS) would facilitate transformation of tyrosine into bisdemethoxycurcumin (ddCCM). The stable expression of this process using this plasmid and a Cas 9 plasmid in Neuro-2a cells can be used to study the effects of curcumin against β-amyloids associated with Alzheimer's disease. Succinylated polyethylenimine has been shown to increase transfection efficiency in serum and therefore can be used to insert these two plasmids into the Neuro-2a cells for analysis. The transfection efficiency of succinylated polyethylenimine in Neuro-2a cells was studied using pGL3 firefly luciferase to determine the optimal polyplex. From this analysis, the 2%, 3% and 6.5% succinylated polyethylenimine demonstrated the best transfection in serum. These polyplexes were then used to transfect the Neuro-2a cells with the TAL, 4CL1 and CUS plasmid. Stable expression was analyzed using gas chromatography. Cell viability was tested using cell titer blue and an ROS assay. The addition of the TAL, 4CL1 and CUS enzymes facilitated production of ddCCM which increased cell viability when exposed to β-amyloids.

638653 A Novel Microfluidic Device to Study Cell to Cell Communication By Dynamic Sampling of Paracrine Factors

Abstract:

Breast cancer is the second leading cause of cancer related death in American women. Breast tumors are composed of a highly diverse cellular make up including cancer cells that interact with the surrounding tumor stroma to produce cancer associated adipocytes, fibroblasts, and macrophages. Secreted factors from these stromal cells modulates the tumor microenvironment (TME) to induce drug resistance and metastasis in cancer cells. There are no commercially available cell co-culture methods capable of the dynamic sampling of extracellular media during an experiment to investigate cell-to-cell communications. To address this, a novel microfluidic device was designed, fabricated, and tested to study intercellular communications during co-culture. The device consists of three parallel fluidic channels connected by an array of microchannels to facilitate the orthogonal diffusion of biomolecules. COMSOL Multiphysics was used to model the mass transfer in the device to optimize channel length and width. The final device design consisted of three parallel 1500 μm long with the two outermost channels for static cell culture and the center channel connected to a peristaltic pump. A proof-of-concept device was fabricated using a combination of soft lithography and polydimethylsiloxane (PDMS) replication. Initial data confirmed the ability to seed cells into the two outer cell culture channels. This microfluidic device will be used in the future to analyze the cell-to-cell communications between MDA-MB-231 breast cancer cells, a triple negative cell line, and adipose-derived stem cells (ASCs) to study how stromal cells enhance the proliferation and drug resistance of cancer cells.

638425 *Computational Fluid Dynamics As an Approach to Ureteroscopy Irrigation*

Abstract:

Phuc Tran[1], Hana Kalčo[1], Rachel Melnyk[2], Andrew Cook[2], Ahmed Ghazi[2], and David G. Foster[1]

[1]Department of Chemical Engineering, University of Rochester

[2]University of Rochester Medical Center

This project focuses on modeling the fluids dynamics during ureteroscopy, the primary surgical treatment for kidney stones. The process consists of retrograde passage of a narrowed lighted scope (ureteroscope), under vision through the urinary channel up to the stones. Ureteroscope fluid irrigation through a channel is essential to pressurize the ureter and the kidney to secure a better vision field during lithotripsy and provide a spontaneous flow passage for kidney stone fragments to move out of the kidney.

Computational Fluid Dynamics (CFD) is a branch of mechanics that creates simulations for a variety of applications and is becoming a useful tool in medical applications to study the physiological flow patterns of fluids in the human body. The aim of this project is to utilize CFD to model and monitor the development of the fluid velocity and intrapelvic pressure during ureteroscopy irrigation. Geometries were obtained from scans of patients' anatomy and inserted into the CFD program. The ureteroscope was placed manually into the geometry at 5 different locations to obtain representative fluid velocity and intrapelvic pressure data that will be validated with known values. Patients' information and ureteroscope's operating conditions are provided by the University of Rochester Medical Center so that the simulation has sufficient boundary conditions to replicate an accurate real-life procedure. The flow behavior of the saline solution within the flow domain is governed by the Navier-Stokes equations. Optimization and manipulation of the geometries allow for the creation of an accurate 3-dimensional mesh throughout the geometries where the CFD software uses finite volume methods and iterative procedures to solve the basic governing equations of fluid dynamics and generate several million data points throughout the geometry.

Through the examination of velocity and pressure profiles over time, an extensive guideline can be established to determine the safe operation time for a specific ureteroscope setting. Our objective is to gauge different patients' specific and unique flow patterns and pressure gradients during the procedure to have a better comprehensive understanding of ureteroscope irrigation.

638247 Enzyme Immobilization within a Hyaluronic Acid Matrix for Biosensor Applications

Abstract:

Enzymes make up an integral system in the success and efficiency of biological reactions; their specific activities have been transferred from biological to synthetic environments and applied within the chemical, environmental, and biomedical industries to increase processing efficiency. There are, however, limitations to enzyme-based technologies including fragility of the biocatalyst, lack in user-controllability of its reaction, ephemeral shelf-life, and low capacity for large-scale usage. To approach such limitations, a technique involving enzyme immobilization within biocompatible and biodegradable hyaluronic acid (HA)-based hydrogels of varying pore-sizes has been explored. Hydrogels are self-assembled upon the respective crosslinking of hydrophobic dodecylamine, hexylamine, and octadecylamine with hydrophilic HA via 1-ethyl-3-[3-(dimethylamino)-propyl] carbodiimide reaction chemistry. With synthesis conducted at 1:10, 1:1, and 10:1 molar ratios of amine to HA, user-controllability of hydrogel surface morphology can be warranted, offering increased applicability of immobilization with a range of enzyme sizes, attributes, and modes of implementation. Upon the production of these amphiphilic matrices, biocatalysts are entrapped within the net-like gel samples for enablement of protection from environment-induced denaturation and degradation while maintaining substrate access to the protein active site and product release from the system, respectively. Diversification of pore sizes in synthesized gels provide trial opportunity for maximization of both concentration of immobilized enzyme as well as activity of such enzyme. Immobilization emphasis is placed on glucose oxidase (GOx), one of the model enzymes used in biosensors and the pharmaceutical industry, before this proof-of-concept immobilization platform can be extended to other enzymes and other applications. Following physical adsorption of GOx within the HA hydrogels, it is envisioned that activity, stability, and sensitivity of the entrapped enzyme can be optimized through manipulation of the platform's pore sizing, as controlled through amine-HA pairing and proportion. Further, attachment of the formulated hydrogels onto functionalized gold electrodes creates opportunity in the development of a biosensor with continuous, real-time tracking capabilities. Such analyses could therefore enhance the body of knowledge and realm of implementation of enzyme immobilization, particularly in the medical sector through applications such as biosensing, drug delivery, and tissue engineering.

638709 Cell Cycle Analysis of Estrogen Receptor Positive Breast Cancer Cells Using Flow Cytometry

Abstract:

In cases of estrogen receptor positive (ER+) breast cancer, the antiestrogen Tamoxifen, a selective estrogen receptor modulator, in its active form (4-hydroxytamoxifen) has been used as a long-term treatment to inhibit tumor growth. However, a large portion of patients develop resistance to tamoxifen overtime by unknown mechanisms. Another breast cancer, Triple Negative Breast Cancer (TNBC), is characterized by the lack of expression of three receptors: human epidermal growth factor receptor 2 (HER2), progesterone receptor (PR), and estrogen receptor alpha (ERa). Since these receptors are commonly targeted in breast cancer, there are limited treatments for TNBC. Our laboratory hypothesized that the accumulation of insulin-like growth factor binding protein 3 (IGFBP-3) contributes to the development of Tamoxifen resistance, and this hypothesis has been supported by previous PCR and immunoblot analysis. In this project, we perform cell cycle analysis on two ER+ cell lines developed to overexpress IGFBP-3. We do this by performing cell cycle synchronization at the G1 or G2 phase using the chemical synchronization drugs Lovastatin and Mevalonic acid. Flow cytometry was used to confirm synchronization. Successful cell cycle arrest and release were observed in the samples treated with Lovastatin and Mevalonic acid, respectively. In future experiments, this protocol will be used to perform direct comparison between triple negative and MCF7 breast cancer cells.

638442 Integration of Microfluidic Washing into Automated NGS Library Preparation

Abstract:

Next Generation Sequencing (NGS) is a critical technology that has massively increased the rate at which genetic material can be sequenced while massively reducing its cost. These benefits have enabled the use of NGS in a wide range of applications, including personalized medicine, agriculture, and archaeology. Preparing samples for NGS is currently a major bottleneck in the overall sequencing workflow, consisting of many pipetting steps and involving small quantities of many reagents which creates room for human error. One critical step in NGS sample preparation is DNA washing, which is carried out multiple times in a typical workflow to remove small DNA fragments from the sample that can negatively impact sequencing. The complexity of DNA washing and overall library preparation makes affordable automation strategies for these workflows very useful.

The goal of this research is to integrate a microfluidic system that enables DNA washing into an existing automated library preparation device. This microfluidic operates by binding a DNA sample to magnetic Solid Phase Reversible Immobilization (SPRI) beads in an input well with a high salt concentration. A magnet is then swept beneath the input well, moving the beads into a washing channel filled with isopropyl alcohol, where smaller unwanted DNA strands are removed from the beads. The beads are then moved into an elution well via magnet, where a low salt concentration causes the longer DNA strands that make up the library to elute off the beads. We have integrated this microfluidic into a library preparation device capable of preparing 8 samples at a time and demonstrated that it can wash libraries after DNA concentration has been increased via PCR amplification. This microfluidic system has been shown to reduce the concentration of unwanted short DNA strands by over 70% relative to an unwashed library while retaining over 51% of the longer DNA strands that are needed for sequencing. This research has demonstrated that this system is a viable method for automated washing of DNA libraries. Further work optimizing the movement and yield of SPRI beads through the IPA channel may lead to further improvements in library purity and yield.

636906 Empirical Modeling of Enteric-Coated Mini-Tablet Dissolution Performance Utilizing X-Ray Computed Tomography and Convolutional Neural Networks

Abstract:

Mini-tablets have been utilized as an alternative to monolithic tablets due to their ease of use for pediatric populations, dose flexibility and tailoring of drug release profiles. Similar to monolithic tablets, mini-tablets can develop film coating and internal defects during manufacturing processes that may adversely affect their dissolution performance. The use of X-ray Microcomputed tomography (XRCT) is well documented for monolithic tablets as a means of identifying internal defects, but applications to mini-tablets have not been well studied. In this study, we have developed a computational program that analyzes reconstructed XRCT images of enteric-coated mini-tablets using deep learning convolutional neural networks to automatically segment individual mini-tablets and quantify individual mini-tablet physical parameters. This algorithm was utilized to determine the correlation between physical features of mini- tablets, such as micro-crack volume and enteric coating thickness, and two-stage dissolution performance in individual mini-tablets. The correlations were then used to predict the two-stage dissolution performance of individual mini-tablets based on the physical parameters obtained from XRCT images. The analysis and results provide insight into the physical variability of mini-tablet populations that are generated during manufacturing, enabling optimization of critical tableting and coating parameters in order to achieve the target dissolution criteria.

High-Throughput Microfluidic Generation and Interrogation of 3D Co-Culture Cancer Spheroids

Sophia Zhou, Anowar H. Khan, Margaret Moe, Elizabeth C. Martin, John A. Pojman, and Adam T. Melvin

Culturing cancer cells in a three-dimensional (3D) environment recapitulates in vivo conditions as these 3D spheroids account for both cell-to-cell contacts and mass transfer limitations. Research in breast cancer therapeutics has shown success in model 2D systems but fail in early clinical trials due to the difference between 2D and 3D cell growth and behavior. Several approaches exist for 3D cell culture; however, droplet microfluidics is a promising approach to generate and interrogate uniform 3D breast cancer spheroids. One area in 3D cell culture that has not been well studied to date is the co-culture of two different cell types in a single spheroid. The paracrine signaling associated with tumor progression between cancer cells and stromal fibroblasts has been suggested as a key contributor to tumor proliferation and drug resistance. Moreover, monocultured cells cannot properly model the tumor microenvironment (TME) which can remodel due to the physical and chemical interactions between cancers cells and fibroblasts. The goal of this work is to utilize a novel approach to grow, interrogate, and evaluate changes in the extracellular matrix (ECM) of 3D co-cultured spheroids. This is accomplished using a thiol-acrylate (TA) hydrogel scaffold that supports cell culture for up to 7 days in a droplet microfluidic trapping array which isolates single droplets in a 300-member trapping array with up to 99% successful trapping. Hydrogel gelation occurs rapidly (~30-40 min) allowing for quick replacement of the oil phase needed to generate the droplets with culture media to support 3D spheroid growth. The device is coupled with a gravity-driven media infusion system to continually supply the spheroids with fresh media to avoid growth stagnation or cell death. Spheroid size, circularity, and diameter were analyzed using custom built MATLAB code scripted to analyze brightfield images of the circular traps. The program masks each trap and isolates the spheroid inside to batch-feed and import data directly into an Excel spreadsheet.

Three different ratios (1:1, 2:1, and 4:1) of co-cultured primary fibroblasts:ER^+ breast cancer MCF-7 cells were studied to determine the role of fibroblast density in increasing spheroid size. Increasing fibroblast concentration correlated to the formation and increase in size of a dark core inside each spheroid. The accumulation of the fibroblasts in the spheroid core was confirmed using CellTracker™, a live stain specific to fibroblasts. The presence of fibroblasts correlated to a significant increase in spheroid area over a 7-day incubation period compared to monocultured MCF-7 cells. To evaluate changes in the ECM, the monoculture and co-cultured spheroids were immunostained for collagen I, an ECM protein thought to play a role drug resistance. Fluorescent imaging showed suggests that monocultured MCF-7 cells secreted less collagen I than fibroblast/cancer co-cultured spheroids. Interesting, a higher amount of collagen I was observed at the core of the spheroids, suggesting the presence of fibroblasts enhances collagen I production in the TME.

638004 Optimization of Microfluidic Cell Encapsulation

Abstract:

Ciara Noelle Smith, Dr. Oberholzer, Dr. Wang, Bruce Zhang

Islet cell transplantation is a type of cell therapy that provides a promising treatment avenue for type-1 diabetic patients. As a minimally invasive procedure, islet transplantation can achieve patient insulin independence. However, the body's immune response can lead to patients reverting to insulin dependence. One way to circumvent this is to encapsulate the islet cells by using alginate, a biomaterial, to form microdroplets which can then be introduced to ionic cross-linking agents (e.g., Ca^{2+}). Here a surface reaction called chelation occurs between the alginate and the divalent cation, forming a coating. Small molecules can cross the coating, like insulin, oxygen, and nutrients. However, larger entities, like immune cells and antibodies cannot cross the barrier.

One method to produce these microcapsules is via a microfluidic device. These devices house microchannels that allow an inlet solution of alginate to be cut into droplets by a secondary, immiscible fluid. The size and surface area of the alginate-based hydrogel are crucial to the kinetic control of insulin release. This technique allows for more precise manipulation of droplet formation, enhances quality control over traditional encapsulation methods, and the ability to better tune the consistency, size, and shape of droplets.

Testing every possible device geometry is not feasible. However, if a dynamic computational model of droplet generation could be built to simulate the microfluidic process, one could then design, test, and improve the microfluidic devices more efficiently.

To address this question, three unique microfluidic devices were first designed using AutoCAD 2019 (Autodesk, Inc.). Then, the designs were processed in the computational fluid dynamics software, ANSYS Fluent 2020 (Ansys, Inc.), and fabricated. The flow ratios between the carrier fluid and the alginate flow were adjusted in a designed experiment, and the effects on size, shape, and droplet formations were used as comparison metrics between the real and simulated behavior.

In this project, different designs were compared to their respective simulated behavior to test whether equivalent flow rate ratios yielded similar flow behaviors at the junctions and similar droplet consistency and shape. Our preliminary results show that actual and simulated flow settings required rates of different magnitudes, and droplet diameters were dissimilar. Analysis of the experimental methodology suggests that fine-tuning the CFD software promises to accurately mimic real microfluidic device behavior with a more robust knowledge of the software and more accurate physical parameters (e.g., surface / interfacial tensions).

638150 Wastewater Monitoring for Sars-Cov-2: Concentration and Normalization

Abstract:

Wastewater based epidemiology (WBE) is an important tool that can be used to gather nearly real time data on the presence of infectious diseases to make important public health decisions. In light of the recent Covid-19 pandemic, researchers are concentrating, extracting, and quantifying the RNA of SAR-CoV-2 in wastewater. Such data can be critical in rapidly evolving situations. To evaluate the extent of infection in a community associated with collected sewage, it may be helpful to normalize concentrations of SARS-CoV-2 wastewater strength. This research has three overarching goals: (1) Compare wastewater concentration methods for samples collected with passive samplers (e.g., tampons); (2) Monitor the spread of SARS-CoV-2 on campus; (3) Evaluate parameters for normalizing concentrations of SARS-CoV-2. For this study, three methods of wastewater concentration were evaluated: Ceres Nanotrap, Concentrating Pipette Select (Innova), and 4S salt-silica. Regarding Goal 2, we monitored a virus that is used as a fecal indicator: Pepper Mild Mottle virus (PMMoV), as well as the following common wastewater constituents: biochemical oxygen demand (COD), total organic carbon (TOC), suspended solids, and conductivity. We collected wastewater, using tampons as passive samplers, every day over 53 days from two manholes at Bucknell University, which serve approximately 300 students, and weekly from a local wastewater treatment plant. Bucknell's wastewater is collected using tampons due to their cost effectiveness. Statistical tests were conducted to compare wastewater concentration methods, and a correlation matrix was created to evaluate the relationship of the virus concentration (PMMoV) with the measured wastewater constituents. All statistical analyses are being done in R Studio.

637037 Development of a Cell Free Homocysteine Biosensor to Assess Folate Deficiency

Abstract:

Folate, commonly known as vitamin B9, is a critical micronutrient in human nutrition that plays an essential role in DNA, RNA, and protein synthesis. Folate deficiency can increase the risk of cancer, cardiovascular disease, neurological disease, and birth defects. Therefore, it is important to have an accurate and prompt method for testing folate levels. In clinical settings, folate levels are measured through testing the concentration in red blood cells (RBC) or serum. Both of these biomarkers are susceptible to false negative results. Additionally, the methods used for testing these biomarkers must be performed in a laboratory using power and expensive equipment, both of which are not easily accessible across the globe. As a result, access to necessary information about one's health is limited to those who can afford it. The Biomarkers of Nutritional Development (BOND) project named folate as one of six micronutrients that warrant immediate attention in biomedical research due to their importance in public health.

Alternatively, plasma homocysteine can be used as a biomarker for folate deficiency, resulting in a new, low-cost, and minimal-equipment method for detecting folate. Homocysteine undergoes a process in the body, remethylation, wherein folate is required to convert it to methionine. Consequently, high homocysteine levels are correlated with low folate levels. Measuring levels of homocysteine in blood plasma is easier than processing RBCs, and leads to more accurate results than serum, which makes this idea very attractive.

We propose a homocysteine biosensor in a cell-free system that uses the interaction between homocysteine and an *Escherichia Coli* transcription factor: MetR. Cell free reactions are becoming increasingly popular in synthetic biology. They allow us to synthesize proteins outside of the cell, which produces faster results, while also being less complicated than whole-cell systems. The cell-free process achieves this by giving us direct access to the inner contents of the cell, removing the barrier of a cell membrane. In our sensor, the homocysteine/MetR interaction drives expression of the reporter gene, Green Fluorescent Protein (GFP), alerting us of the patient's homocysteine levels. Through extensive testing we have proven that our proposed biosensor detects homocysteine concentrations within the physiologically relevant range.

638268 Cold Triggered Drug Release from Polymersomes

Abstract:

Cold triggered drug release stands to be more advantageous due to lower cost and greater accessibility. The goal of this project is to develop a cold triggered drug delivery system. The platform has 3 elements: the drug, cold therapy, and thermo-responsive polymersomes. To model this type of platform, three methods were tested. In the first method, a polymer was mixed with a fluorescein solution, cooled, and then put in a dialysis tube inside of a centrifuge tube with 14 mL of 37 C Phosphate Buffer solution (PBS), which was exchanged at given time intervals. At each time interval, the amount of drug released was measured as a cumulative percentage calculated to better compare the cold triggered intervals with those at the standard 37 C. These percentages were found by obtaining concentration values from an average of relative absorbances. Another method tested this with direct injection of fluorescein solution into a sample vial, no longer requiring a dialysis tube. The measurements were taken in the same manner but from the supernatant of the mixture, only replacing that volume of PBS at each interval. The final method was also with direct injection, into a 24 wells plate, yet differed from the other two methods by measuring the relative fluorescence instead of absorbance. The results from the fluorescence method proved to be the most promising, showing a steady cumulative release over time that was noticeably increased at the cold triggered intervals. Future testing is necessary for the long term work yet this shows great possibility in the application of thermo-responsive polymers. The significance of this finding is to develop a platform to remotely trigger drug release on demand.

637572 Targeted Delivery to the Epidermis Via Angled Insertion of Microneedles

Abstract:

Peanut allergies affect 1 - 2% of adults and 4 - 8% of children. A traditional method for treating peanut allergies is by allergy shots, formally known as subcutaneous allergen immunotherapy (SCIT). This approach is daunting for young children as they are often afraid of injection by needle. Also, SCIT with antigens such as *Arachis hypogaea 2* (*Ara h 2*), the dominant peanut allergen, can result in severe allergic immune responses—in some cases, even anaphylaxis. One of the main reasons for these adverse effects is rapid clearance of antigens by macrophages in the dermis and subcutaneous layers of the skin.

The safety and efficacy of targeting allergen-specific immunotherapy to the skin can be augmented by increasing the localization of the antigen in the epidermis, therefore decreasing the amount localized in the dermis/subcutaneous layers. Microneedles (MNs) have a geometry that allows for antigen delivery targeted to epidermal thickness (30 - 140 μm). They also provide a solution to the pain associated with SCIT. In addition, MNs circumvent patient compliance issues caused by needle phobia.

This study employs angled insertion of 250 μm long MNs (coated with *Ara h 2*-containing solution) into the skin for the purpose of maximizing antigen localization in the epidermis. Three angles of insertion (relative to the plane of the skin) were studied: 90° (straight insertion), 45°, and 20°. Stainless steel MN arrays are first dip-coated on a micron-scale with a fluorescent *Ara h 2* solution. Angled (45° and 20°) MN insertions were executed using a biplanar, 3D-printed device with an adjustable angle of elevation while straight insertions were performed by hand. These pieces of skin were then histologically sectioned, and the insertion sites were imaged with red fluorescence under a microscope. Finally, the sections were H&E stained and imaged under a bright-field microscope to differentiate the layers of the skin.

These data show that angled insertion was successful to limit MN insertion depth to facilitate targeted delivery to epidermis. We also include fluorescent and companion H&E-stained images of representative insertion sites, which together give insight on the extent of epidermal targeting. Opportunities for further study include *in vivo* insertions on porcine/murine subjects and immunohistochemistry to measure specific antigen localization in the epidermis.

638421 Encapsulation of Retinoic Acid and Amitriptyline Hydrochloride By Pluronic F127 Using Sonication and Solvation Dry Film Method

Abstract:

In the recent decade, strokes have left many individuals with damaged nervous systems, and left especially elderly with limited motor function, as strokes often result in the loss of GABAergic neurons, which are a crucial part of the brain's communication pathways. A possible method of recovery would be to use neural stem cells and supply them with small molecules to direct neural stem cell (NSCs) differentiation. However, two effective molecules, retinoic acid (RA) and amitriptyline hydrochloride (AH) are hydrophobic, making drug delivery difficult. Nanocarrier encapsulation has been an effective strategy to deliver hydrophobic drugs, and thus here F-127 micelles have been chosen as the delivery system. This was done by using sonication to ensure the drugs are encapsulated, as well as dry film solvent methods ideally to evenly distribute the drugs among the nanocarrier. DLS measurements showed inconsistent sizes of micelles across both methods, by dry-film solvation appear more consistent in making small micelles (28-32nm diameter), along with some larger micelles (200-400nm diameter). TEM pictures confirmed the construction of these micelles, but quantitative data is still needed to confirm the drugs were successfully encapsulated.

638452 Production and Stabilization of the Therapeutic Peptide Angiotensin 1-7 for Oral Delivery

Abstract:

Angiotensin (Ang) 1-7 is a peptide in the renin-angiotensin system that regulates the cardiovascular and metabolic systems, making it an attractive therapeutic target for a number of disorders. However, Ang 1-7 has a short half-life *in vivo*, decreasing the efficacy of this therapeutic. As shown with small molecules and proteins, polymer carriers prolong the *in vivo* stability of therapeutics. Ang 1-7 has a net cationic charge, rendering it amenable to electrostatic encapsulation with anionic polymers. The goal of this research is to scale peptide production and study the electrostatic encapsulation of Ang 1-7 within polymers. In order to produce Ang 1-7 at appropriate scales, we improved the purification process for Ang 1-7 by modifying both analytical and preparative high-performance liquid chromatography (HPLC) elution profiles. These modifications reduced the time for a single preparative HPLC batch from 1 hour to 20 minutes, saving over 1 L of solvent. With injections of 30 mg crude Ang 1-7 per batch, we consistently recovered 65% of the peptide. For encapsulation of Ang 1-7, we also performed reversible addition-fragmentation chain-transfer (RAFT) polymerization to synthesize an anionic poly(methacrylic acid) suitable for Ang1-7 encapsulation. We confirmed the polymer structure with nuclear magnetic resonance (NMR) spectroscopy and determined the molecular weight by gel permeation chromatography (GPC). This poster describes preliminary encapsulation studies of Ang 1-7 in poly(acrylic acid) and poly(methacrylic acid) using dialysis and HPLC. These results lay the foundation for future development of an oral delivery system for Ang 1-7.

636931 Design of Novel Tumor-Binding Cytokine Immunotherapies for the Intratumoral (IT) Treatment of Solid Tumors

Abstract:

Conventional therapies to treat solid tumors can inflict many complications on the patient's body due to a lack of tumor-killing specificity. To avoid this risk, cancer immunotherapy that activates the patient's own immune system to selectively fight cancer is a better alternative. One such immunostimulant currently being investigated in clinical trials is a type of cytokine called GMCSF (granulocyte macrophage-colony-stimulating factor). Local injection of GMCSF can recruit and activate antigen-presenting cells (dendritic cells) into the tumor lesions, thereby initiating anti-tumor responses. However, the main limitation of local GMCSF treatment is its lack of retention within the tumor tissue leading to systemic toxicity in patients. In order to address this problem, this research focuses on the design of an engineered-GMCSF (eGMCSF) that exhibits superior retention and immune stimulation at the injection site, thereby providing a safer and more efficacious treatment strategy for solid tumors. This is attained by fusing GMCSF with a tumor binding peptide that is designed to bind to tumor ECM (extracellular matrix) components, such as hyaluronic acid (HA) and Fibronectin. The efficacy of the eGMCSF is investigated by testing its tumor binding capabilities using an *in vitro* tumor model. Additionally, the immunological capabilities of eGMCSF are studied by elucidating the proliferative effect on dendritic cells (DCs).

636444 Optimizing the Crystallization and Coating of Drugs for Regenerative Medicine

Abstract:

Regenerative medicine is an emerging field that uses engineered biomaterials in tandem with the human body's natural healing process to regenerate damaged tissue and organs. Recent studies on Murphy Roths Large (MRL) mice demonstrated their capability to regenerate injured tissue due to an overexpression of hypoxia-inducible factor-1α (HIF-1α) throughout the initial stages of the wound healing process. A therapeutic strategy to beneficial upregulation of HIF-1α is through inhibition of prolyl hydroxylase (PHD), an enzyme that regulates HIF-1α levels through hydroxylation of proline residues, leading to degradation of HIF-1α. Recently we reported the use of a small molecule drug inhibitor of PHD for tissue regeneration. Since this class of drugs are poorly soluble, development of new formulations may open the door to further enhancements in regenerative medicine. The objective of this study is to enhance the efficacy of a PHD inhibitor through crystal surface engineering. In particular, we are interested in surface modifications that increase the solubility of insoluble drugs. One approach is to incorporate mussel-inspired polymers, like polydopamine (pDA) coating on nanoparticles, to improve drug targeting, residence time and bioavailability. Therefore, in this study, the drug release and encapsulation efficiency of pDA-coated PHD inhibitor nanocrystals were thoroughly investigated. Drug crystal fabrication was performed using flash nanoprecipitation, where the modification of preparation conditions were employed to control crystal size. For example, by changing the aqueous to organic ratio of crystallization solvent, mixing and cooling temperatures, crystal size decreased by 93%. Additional optimization of the pDA coating was performed. The pDA-coated drug crystals were analyzed by TEM and drug release studies were conducted with HPLC. This innovative approach leverages both potential pro-regenerative drugs with bio-inspired coatings to facilitate the development of next-generation drug targeting and tissue regeneration.

638548 Cellular Response to Growth Factors and Doxorubicin of Breast Cancer Cells with Different Insulin Receptor Isoform

Abstract:

Insulin receptor (IR) plays a crucial role in cellular growth and proliferation through the regulation of metabolism. IR-A and IR-B isoforms are expressed by differential splicing of the IR gene, where IR-A modulates proliferation through inducing growth signaling factors, while IR-B signals primarily for metabolic pathways. Therefore, overexpression of IR-A in humans can lead to tumorigenic activity, which can be observed in endocrine-resistant breast cancers. IR also interacts with Insulin-like Growth Factors (IGF) along with IGF receptors in the tyrosine kinase receptor family.

In this study, breast cancer cell lines derived from MCF-7L with various IR isoform phenotypes were characterized through their response to various growth factors and resistance to doxorubicin, a common chemotherapy drug, measured by MTT assay. Here, we demonstrated a synergistic effect of the two IR isoforms in response to the ligands and in developing resistance to chemotherapy drugs. Furthermore, estradiol and growth ligands (Insulin and IGF-I) together provide a greater synergistic effect in MCF-7L cells compared to IR-A and IR-B cells. Further investigation will include co-infection of tagged IR-A and IR-B isoforms without endogenous IR expression to compare with the parent MCF-7L cell line, and their dose response to growth factors tested in this project.

638120 Characterizing the Kinetic Behavior of Candidate Threonine Transaldolases

Abstract:

Enzymes are appealing catalysts, driving specific chemical reactions under ambient conditions in aqueous environments. Upon incorporation into protein sequences, non-standard amino acids (nsAAs) can imbue enzymes with novel chemistry, unlocking avenues for biocatalysis and pharmaceutical applications. Not only can nsAAs greatly contribute to the development of innovative protein technologies, but some nsAAs have pharmacological relevance themselves. Threonine transaldolases (TTAs), which comprise a recently characterized enzyme class, condense an aldehyde with threonine to synthesize nsAAs with β-hydroxy groups. These enzymes have great potential for β-hydroxy-nsAA synthesis given their high enantioselectivity and thermodynamic favorability in biocatalysis, but only one TTA (ObaG from *Pseudomonas fluorescens*) has been studied closely in literature. My research seeks to characterize the substrate specificity and kinetic behavior of other candidate TTAs. Using a sequence similarity network constructed around ObaG, our group selected eight sequences to probe a diverse spread of potential TTA chemistries. Through *in vitro* analysis with a coupled-enzyme assay, I verified that one of the TTA enzyme candidates I cloned, expressed, and purified demonstrates TTA chemistry. Compared to ObaG and three other TTAs previously-confirmed by our group, this new TTA has the lowest K_M for threonine when using the model substrate phenylacetaldehyde. This finding indicates that the new TTA has an enhanced affinity for its threonine donor—a metric initially reported as a hurdle for ObaG kinetics.

637925 Gelatin Methacrylate Hydrogel Suitability for Soft Polymer Glioblastoma Cell Culture

Abstract:

Glioblastoma multiforme (GBM) is an aggressive, deadly form of brain cancer with a prognosis of approximately 12-15 months. This bleak prognosis can be attributed to the ineffectiveness of current GBM treatment methods, which can be slow, painful, and expensive for patients. Advanced tissue and polymer engineering provide alternative avenues to improve GBM research, treatment, and therapy through the enhancement of organoid technology. In short, organoids are miniature organs grown in a dish and developed from stem cells that are induced into differentiation, typically through mechanical signaling from an extracellular scaffold such as a hydrogel. Conventionally, GBM stem cells are grown *in vitro* using tissue culture flasks made of polystyrene (TCPS). In this form, polystyrene is a hard plastic with a stiffness seven orders of magnitude greater than that of human brain tissue. Additionally, TCPS does not accurately depict the highly heterogenous tumor microenvironment (TME) of GBM. This project explores the mechanical properties of gelatin methacrylate (GelMA) and its compatibility with GBM organoids. GelMA was selected due to its cost efficiency, biocompatibility, and controllable stiffness. Samples of 3%, 5%, and 7% weight/volume GelMA hydrogels were synthesized using 3D-bioprinting techniques. Samples were mechanically characterized using various methods, including enzymatic degradation tests, fluid swell tests, and rheological testing, to determine which weight/volume percent exhibited a stiffness most similar to that of human brain tissue. Based on rheological testing, 5% weight/volume GelMA hydrogel was most similar to native brain tissue and is likely the best candidate for *in vitro* culture of GBM tumor cells in the presence of soft biopolymers. In conclusion, soft polymers like GelMA can closely model tumor microenvironments and improve cell culture conditions to further advance GBM organoid technology.

637959 Therapeutic Potential of Lentivirus-Mediated shRNA Delivery in Spinal Cord Injury

Abstract:

Traumatic spinal cord injury (SCI) results in an initial primary injury, followed by secondary events that exacerbate the damage. As a key factor of secondary injury, inflammation is initiated in part by the rapid influx of circulating immune cells. These cells begin to release various inhibitory factors, which can contribute to additional neuronal cell death, demyelination, and functional deficits following SCI. P38 mitogen-activated protein kinase (MAPK) is specifically activated after SCI, which results in the production of multiple factors that inhibit axon growth and activate nociceptors, contributing to various functional deficits. Short hairpin RNA (shRNA) interference is an efficient tool for the specific knockdown of disease-related gene expression after SCI. In this study, P38 MAPK-shRNA encoded-lentiviral vectors (LV-shP38) were delivered into the injured spinal cord to silence downstream P38 MAPK signaling. We performed qRT-PCR to investigate the effects of LV-shP38 from the spinal cord, dorsal root ganglia, and spleen. The data indicate that LV-shP38 significantly increases the expression of pro-regenerative factors, while reducing the expression of pro-inflammatory factors compared to control group (luciferase encoded lentiviral vector). Our LV-mediated gene silencing therapy will enable efficient and extended delivery of shRNA, thereby overcoming current hurdles to successfully introduce RNA interference-based therapies into the clinic.

638609 Effect of Shear Stress on Secondary Metabolite Production in Taxus Plant Cell Culture

Abstract:

Cancer is the second most common cause of death in many developed countries, making development of cancer treatments highly important. Paclitaxel is an FDA approved drug used for treatment of ovarian, breast, and lung cancer, as well as Kaposi's sarcoma. Due to the high demand for paclitaxel, it is important to have a robust and scalable production method. Plant cell culture (PCC) of *Taxus* species is the current state-of-the-art for production of paclitaxel on an industrial scale.

One major drawback of producing paclitaxel using plant cell culture is low and unstable yields. Thus, exploration of methods for increasing yield of paclitaxel in *Taxus* PCC is an ongoing field of research. When plant cells divide, they often form masses of cells called aggregates that remain connected through their cell walls. Previous research in the Roberts laboratory has shown that large cellular aggregates produce less paclitaxel than smaller aggregates, possibly due to decreased shear force. To this end, we sought to investigate mechanical shearing of aggregates as a method for decreasing average aggregate size, which could be a method of increasing paclitaxel production. The two aims of this project are the following: to determine how shear stress affects *Taxus* secondary metabolism, and to design a device for continual shearing of plant cell cultures.

Mechanical shearing had an inconclusive effect on aggregate size distribution, with only sheared, methyl jasmonate elicited cells having aggregates that decreased in size over time. All other treatments did not lead to a decrease in aggregate size distribution over time. Taxane quantification data showed that a combination of shearing and methyl jasmonate elicitation yielded the highest concentration of paclitaxel. Secondary metabolites, which are chemical compounds that are produced in response to stress, also increased with both shearing and elicitation. This suggests that shearing does not specifically upregulate production of taxanes, but more generally upregulates secondary metabolism as a whole. Finally, a continual shearing device was designed with the intent to later construct and study this device as an alternative to manual shearing with a pipette

638612 Extraction of Pectic Oligosaccharides from Cranberry Pomace

Abstract:

Naturally occurring polysaccharides have been of great interest for applications in the food and biomedical industry owing to their ability to self-assemble into gels with the help of different cross-linkers. Cranberry Pomace has been shown to contain pectic oligosaccharides, which are composed of a branch of sugars called xyloglucan. Xyloglucan is a structural oligosaccharide, found primarily in plant cell walls, with a basic structure of a glucose backbone chain with alternating xylose residues attached to every first and fourth or sixth glucose subunit. Most xyloglucan oligosaccharides contain more than just one type of residue in their structure such as galactose, arabinose, rhamnose, etc. which provides an overall hydrophobic and/or hydrophilic oligosaccharide.

Thermoresponsive gelation is achieved by altering the hydrogen bonding between the main chain and side chain with the help of physical cross-linking. However, in order to modulate the sol-gel temperature, the side chain residues in structure play a vital role in controlling the hydrophilicity. Here, we have used a sequential extraction strategy to isolate oligosaccharides from cranberry pomace. We started with obtaining the alcohol insoluble solids (AIS) and followed a series of extraction steps involving solutions with known ionic strength to remove undesired oils and acids to isolate oligosaccharides at the end. 1H NMR spectroscopy was used for qualitative characterization of the final product.

638374 Studying the Effect of Cellulase Supercharging on Cellulosic Biomass Hydrolysis

Abstract:

Lignocellulosic biomass, composed of cellulose, hemicellulose and lignin, is broken down to simple sugars in nature by enzyme cocktails secreted by various microbes. Glucose, produced by the breakdown of cellulose mediated by cellulase enzymes, can be fermented to ethanol and used as transportation fuel. However, the cost of enzymes is a bottleneck limiting wider commercial adoption of cellulosic biofuels. Non-productive binding of enzymes to lignin and cellulose, is proposed to be a key deterrent to cellulose hydrolysis. Recently, we had shown that supercharging cellulase enzymes through modification of surface residues is an effective strategy to reduce inhibition of cellulase activity by lignin (*Whitehead et al., ACS Sustainable Chem. Eng. 2017, 5, 7, 6247–6252*). However, supercharged enzymes in this study showed a significant reduction in activity compared to wild-type enzymes despite desirable performance in terms of lignin inhibition. To address this gap in our knowledge of how supercharged enzymes function, we designed a library of mutants for carbohydrate-binding module (CBM) family 2 proteins found naturally associated with an endocellulase Cel5A (from *Thermobifida fusca*). These mutant CBMs were attached to GFP reporter fuorescent protein to study binding to cellulose and lignin. In addition, the mutants were fused with wild-type Cel5A (*T. fusca*) catalytic domain to understand the impact of CBM supercharging on enzyme hydrolytic activity on pure cellulose and pretreated lignocellulosic biomass substrates. Here, we report our computational design strategy for generation of the mutant library and binding/activity results which present a comprehensive picture of protein supercharging as a potential strategy to alleviate enzyme non-productive binding to cellulosic biomass.

638231 High Temperature Measurements of Multicomponent Foods

Abstract:

Low water activity and low moisture foods are generally considered safe due to the inability of microorganisms to grow under such conditions. Recently, there have been a surprising number of recalls on low water activity foods due to pathogens such as *Salmonella agona*. To understand this issue, we investigated the dependence of water activity on *S. agona* inactivation in multicomponent (carbohydrate, protein, fat) food systems. Initial results indicated that a high fat multicomponent food requires a longer time for complete *S. agona* inactivation. Whereas a high carbohydrate or high protein multicomponent composition would require a shorter time for the completion of *S. agona* inactivation. We conducted a follow on study to confirm water activity at high temperatures and the starting moisture contents of each multicomponent food system using a series of moisture content and water activity measurements. To date, the initial and current data has been consistent across the studies. These results imply that further studies are required to understand how different types of carbohydrates, proteins, and fats will differ while undergoing *S. agona* inactivation. These further studies may be looked into by using various different carbohydrate, protein, fats compounds and structures that may affect the *S. agona* inactivation significantly, on structure alone or combinations of various different structures.

636656 Enabling the *De Novo* Production of Psilocybin Using an *E.coli* Co-Culture Approach

Abstract:

Psilocybin is the psychoactive prodrug of psilocin that has been shown to lessen the negative effects of PTSD, depression, and addiction. Biosynthetic production of psilocybin has been previously reported in bacteria, however, this was achieved through the metered supplementation of the starting substrate, 4-hydroxyindole. Here, we demonstrate the first example of *de novo* biosynthesis of this important molecule in a bacterial host, with all carbon deriving solely from glucose.

De novo synthesis is the process of producing complex biomolecules from simple carbon sources without the addition of supplements or precursors. To enable this process, we used a P450 monooxygenase, psiH, for the selective hydroxylation at the 4-position on tryptamine, resulting in 4-hydroxytryptamine. This intermediate product was then phosphorylated by the kinase, psiK, and dimethylated by a SAM-dependent methyltransferase, psiM, resulting in psilocybin. In order to maintain a redox balance, a cytochrome P450 reductase (CPR) was coupled with psiH; this required screening of multiple CPRs and gene orientations. Once we determined the top producing strain from genetic optimization efforts, we explored a range of fermentation conditions, including induction times, IPTG concentrations, and iron concentrations. Since P450s are heme proteins, they have an iron-porphyrin group. This need for iron acted as motivation to test different iron salts with our media to allow the P450 to fold correctly and carry out its catalytic processes more efficiently. These genetic and fermentation optimizations were performed in a high throughput well-plate format and top strains and conditions were further scaled to shake flask and benchtop bioreactors enabling enhanced pathway performance. To enable rapid evaluation of P450/CPR combinations, a co-culture approach was used in combination with the previously identified psilocybin producing strain, pSilo16, capable of psilocybin production from 4-hydroxyindole.

638385 Ethylene Synthesis By Genetically Modified *Rhodococcus Opacus* PD630 on Synthetic Human Urine

Abstract:

ETHYLENE SYNTHESIS BY GENETICALLY MODIFIED *RHODOCOCCUS OPACUS* PD630 ON SYNTHETIC HUMAN URINE[1]

A. Williams and K. Wawrousek

Deep space exploration necessitates the ability to generate fuels, chemicals, food, and pharmaceuticals off-earth with limited resources, as it would not be feasible to send all potentially necessary materials via spacecraft. To advance toward the goal of engineering microbes to produce materials that may be needed for human life on other planets such as Mars, we are genetically modifying the non-model bacterium *Rhodococcus opacus PD630* to produce ethylene, the monomer necessary for polyethylene plastic production. *R. opacus PD630* was chosen because this genetically tractable, oleaginous microbe is known for its metabolic plasticity, so it may be able to grow on several organic waste streams. To utilize a nutrient source available off-Earth, we first tested this bacterium for growth on synthetic urine. Monitoring the growth behavior of the bacterium, growth curve studies on Luria-Bertani Broth and synthetic human urine were conducted on *R. opacus* PD630, quantifying overall culture growth by the change in optical density at 600 nm. Within the synthetic urine growth media, our bacterium entered the stationary phase after 50 hours of incubation at 30°C and shaking at 150 RPM, reaching a final optical density of 1.3. To engineer *R. opacus* PD630 to produce ethylene, the ethylene forming gene (efe) was codon optimized for *Rhodococcus* and introduced into the pBAV1K plasmid for expression under the constitutive PsbA promoter. Gene expression will be confirmed via immunoblot, and ethylene production will be measured with gas chromatography, as ethylene is a gas at the 30°C growth temperature.

Additional Key Words: *Rhodococcus opacus*, genetic modification, ethylene production, human urine, Mars colonization, efe gene.

[1] To be presented at the University of Wyoming's Undergraduate Research Day in December 2021 via poster and oral presentation.

[2] Karen Wawrousek, Ph.D. is an Assistant Professor in the Department of Chemical Engineering, University of Wyoming, Laramie WY 82071. Alexandria Williams is an Undergraduate Senior Chemical Engineering Student, Dept. of Chemical Engineering, University of Wyoming, pursuing an MS in Biopharmaceutical Process Engineering.

638493 Effects of Mixing on Volatile Fatty Acid Production in Anaerobic Digestion

Abstract:

Anaerobic digestion(AD) is a common method of waste treatment where organic waste is recycled into a variety of products using microorganisms in the absence of oxygen. Contemporary AD processes are typically optimized for Methane for use in combustion reactions. Other value-added products of anaerobic digestion include VFAs (Volatile Fatty Acids), which can be extracted and used in biofuels. This investigation focused on determining the effects of agitation on VFA production in AD. Possible effects include suspending solids in the reactor, enzyme-solid attachment, and acetogen-methanogen syntrophy. Results from experiments with four different agitation rates suggested that "middle ground" agitation promoted VFA production. The lowest (50RPM) and highest (800 RPM) settings showed slower VFA production in the first few days than did the moderate (70RPM and 400RPM) settings, suggesting that the hydrolysis step in AD was limited. It is speculated that lower RPM agitation contributes to solid layer formations, leading to limited surface area for food waste hydrolysis. However, too high an agitation rate may inhibit enzymes from attaching to food waste solids, limiting hydrolysis. A promising area for future research could include isolating each step of the anaerobic digestion process to further quantify the effects of agitation on specific stages of AD.

636558 Techno-Economic Assessment of a Bioprocess for Long-Chain Dicarboxylic Acid Production from Vegetable Oils: A Case Study for Distillers Corn Oil

Abstract:

This study analyzes the techno-economic feasibility of a bioprocess for production of unsaturated long-chain dicarboxylic acids (LCDCAs) from distillers corn oil (DCO). Currently, there are a limited number of LCDCA products readily available, with dodecanedioic acid and brassylic acid (C12 and C13 LCDCAs) being the most widely marketed products of this type. This process could allow for the expansion of the LCDCA and specialty polymer market by introducing longer LCDCAs which can further tune polymer properties, fill performance gaps, and offer unique properties such as crosslinking and adding additional functional groups. A conceptual design of the process was developed based on conversion of the triglycerides in DCO to fatty acid methyl esters (FAMEs), followed by bioconversion of FAMEs to LCDCAs. The upstream portion starts with a transesterification reactor where DCO is fed into a batch reactor with excess methanol and sodium hydroxide catalyst. Then, the effluent consists of FAMEs, glycerol, and excess methanol. The effluent is distilled to yield FAMEs and glycerol for fermentation and the excess methanol is recycled back into the transesterification reactor. The fermentation section of the process involves multiple bioreactors which utilize the *Candida viswanathii* yeast strain to convert the FAMEs to LCDCAs. The downstream portion of the process contains multiple separation and filtration units which remove any impurities from the fermentation and upstream portions of the process. The resulting product is then distilled twice using vacuum distillation to recycle any unreacted FAMEs into the fermentation portion of the process and to separate the 98.0% pure C-18 LCDCA product from the 48.5% pure C-16 LCDCA product. Sensitivity analysis was employed to estimate the minimum selling price for the LCDCA product with respect to variation in feedstock cost and fermentation time. Within the historic range of DCO prices, feasible scenarios exist, yielding a selling price of LCDCA less than $5.50/kg, competitive with the price of available specialty dicarboxylic acids such as sebacic and dodecanedioic acid. Alternative feedstock and processing scenarios were also explored, as the process could be adapted to utilize other vegetable oils or available fatty acids such as oleic acid and include unit operations such as hydrogenation reactors to yield a fully saturated LCDCA product.

638139 Organic, Regenerative, and Conventional Cultivation Strategies, a Comparative Vitamin C Analysis in Tomatoes.

Abstract:

In this study, different varieties of grape and full-size tomatoes were used to compare their vitamin C content depending on their method of cultivation. These methods included conventional supermarket varieties, organic supermarket varieties, local non-organic varieties, and varieties from a farm practicing a regenerative-organic approach to cultivation. The supermarket varieties selected were grape and hothouse nonorganic tomatoes, and organic grape tomatoes. Hothouse and grape tomatoes were selected from local non-organic sources, and these were also selected from the local regenerative farm. These were all kept in refrigerated storage for less than 24 hours to minimize analyte degradation before processing. To begin processing, the tomatoes were washed, dried, had their stems and any brown spots removed, and then were weighed. Following this they would be cut up and homogenized with a hand mixer, then the tomato pulp would be diluted twenty five percent by volume with water to begin to lower the concentration of L-Ascorbic Acid (L-AA) to within the sensitivity range of the High Performance Liquid Chromatography (HPLC) column. After dilution the pulp would be transferred to plastic vials for centrifugation and subsequent filtration to remove any remaining tomato solids which might interfere with the operation of the HPLC column. The extract would then be transferred to amber HPLC vials, which would help to reduce the impact of UV light on the degradation of L-AA in the tomato extract. Standards of L-AA in the expected range of concentrations for L-AA in the tomato extract would be prepared using serial dilutions to be measured with each trial in the HPLC column, these also would be stored in amber vials. Once processing began, it proceeded quickly to preserve the integrity of the analyte in the tomato extract and the standards of L-AA standards used in HPLC testing. From comparing relative concentrations of vitamin C between each method of cultivation, we will have further insight into the relationship between method of cultivation and nutritional profile of tomatoes.

637737 The Effectiveness of Solid-Phase Antimicrobials Against Gram-Positive Bacteria in Food Processing

Abstract:

Contamination of food products during production remains an immense risk to public health with foodborne pathogens causing outbreaks each year. The dairy industry is searching for methods to eliminate or mitigate bacterial growth, most prominently that of *Listeria monocytogenes* (LM), a gram-positive bacterium. LM infection is particularly concerning due to the alarmingly high mortality rate of 20-30% for young, elderly, and immuno-compromised individuals. In addition, consumer demands for lower levels of preservatives incentivize the discovery of novel antibacterial methods. Solid-phase antimicrobial surfaces represent one such method of targeting and destroying these gram-positive bacteria.

The feed spacer material of a spiral-wound membrane element represents an ideal surface to functionalize with a solid-phase antimicrobial agent. The feed spacer is composed of polypropylene (PP) and serves to induce turbulence and preserve flux. Furthermore, the spacer is in direct contract with the dairy process stream and can be implemented throughout the manufacturing process.

Compounds shown to possess antimicrobial properties in solution were selected to attach to the PP membrane to confirm any suspected solid-phase antimicrobial activity. Phenols and polyphenols were chosen as they have been shown to inhibit bacterial cell enzymes and disrupt their phospholipid bilayer leading to cell lysis. Quaternary ammonium compounds (QACs) were also chosen as their positive head and hydrophobic tail have been shown to heavily disrupt the integrity of the bacterial cell wall leading to cell death.

The surface chemistry of the PP feed spacer was altered with phenolic and quaternary ammonium compounds to confirm any solid-phase antimicrobial activity. Gram-positive bacteria served as surrogates to LM due to their similar cell wall structures and ease of use. The functionalized PP mesh was qualitatively assessed by dye characterization, with the absorption of dye confirming the successful alteration of the PP mesh. Any antimicrobial activity of phenolic modified mesh, represented by eugenol and tyrosine, was found to be negligible. This is suspected to be a dosing issue or due to phenols/polyphenols antibacterial method of action being inapplicable in the solid phase. The QAC altered mesh, however, was found to eliminate gram-positive bacteria below detection limits. These results suggest that PP membranes functionalized with a QAC could serve as an extremely effective antimicrobial surface, providing a novel solution against pathogenic bacteria in the dairy industry and food processing as a whole.

638675 Exploring Intermolecular Drug-Nanocarrier Interactions for High Drug Loading

Abstract:

Introduction: Nanomedicines offer distinct advantages over traditional small-molecular drugs such as improved
tropism and reduced side-effects. Despite this, nanomedicines have not lived up to expectations, with relatively few
clinical approvals. One contributing factor for this is carrier immunogenicity, which limits the total dose. It has
become recently appreciated that higher drug-loading (wt% drug in NP) leads to higher maximum tolerated doses,
and improved outcomes. Here, we have developed a polymeric nanocarrier library where the hydrophobic 'core'
contains varied proportions of H-bonding (monomer A) and π-interacting (monomer B) moieties so to optimize
drug-nanocarrier interactions.
Materials and Methods: Hydrophobic polymer 'cores' were synthesized via anionic ring-opening polymerization of monomer A and monomer B at 100, 90, 80, 60, 50 40 and 30 wt.% of B and end-capped with an
alkyne moiety. The core polymers were conjugated to hydrophilic stealth corona block, (poly(dimethylacrylamide)
(PDMA)-azide) via the copper-catalyzed Huisgen click-reaction and purified by dialysis. Difficult to load drugs, containing a mixture of H-bond acceptors/donors and π-interacting groups, were chosen. Polymer and drug were dissolved at a 1:1 ratio and solvent was removedby nitrogen-flow, followed by high vacuum. Films were rehyd rated at 10 mg/mL drug in DI water (stirred at 1200
rpm / 50°C) then centrifuged for 4 minutes at 7000 rpm; the supernatant was separated from pellet and
loaded drug was determined by HPLC analysis. The resulting concentration was used to calculate loading efficiency
(LE) and load capacity (LC); the latter was determined gravimetrically after supernatant lyophilization.
Results and Discussion: Drugs that have proportionally large potential for both H-bond and π-π interactions were
expected to load well into polymeric cores with complimentary interactions. Paclitaxel, which contains both H-bonding and π-interacting moieties displayed a loading maximum at 90%B (i.e., high π-interactions + low H-bond).
In contrast, etoposide displayed significantly higher (>2x) LE+LC in the absence of any H-bonding groups in the
core (100%B), whereas bortezomib and cyclosporin A (two drugs with a predominance of strong H-bond accepting
amide groups) had an increased loading with increasing H-bonding.
Conclusions: Highest drug loading occurs when the H-bond and π-π interactions between the core polymer and the
drug are maximized; for every drug this likely needs to be tailored. 'Ultra'-high loading (>35% LC) of paclitaxel
and etoposide was achieved after core optimization, significantly higher than benchmark controls.

Future works will evaluate additional drugs as well as new core and corona chemistries to gain better understanding of the structure function relationships of drug-loading.

638326 Quantifying the Capture of Dye-Loaded Silica Nanoparticles on Magnetic Beads

Abstract:

The current standard method for detecting proteins and antibodies is an enzyme-linked immunosorbent assay, or ELISA. Even though this method proves to be successful, the procedure is very complex and time consuming. An alternative method that is demonstrated in this study is to use dye-loaded silica nanoparticles coupled with a sandwiching immunoassay approach to detect proteins. We hypothesize that high surface area, dye-loaded silica nanoparticles will provide a rapid and robust detection of proteins and antibodies with high signal to noise ratio. The binding of biotinylated fluorescently labeled silica nanoparticles to streptavidin (protein)-functionalized magnetic beads served as a model system to investigate the variables relevant to signal detection from dye-loaded nanoparticles, where the strong binding of biotin-streptavidin binding is commonly used to link biomolecules to surfaces. Biotin was covalently attached to the surface of amine-functionalized nanoparticles. Two types of dye-loaded biotinylated mesoporous silica were investigated (fluorescein isothiocyanate (FITC) dye bound through covalent attachment to surface amines or sulforhodamine (SR)-101 dye bound through surface charge to surface amines). To amplify the signal of these fluorescently labeled nanoparticles after capture on the magnetic beads, a high pH buffer was introduced to the solution. The fluorescence intensity of the dye released from the bound particles was measured via spectroscopy as a function of time and particle concentration. The biotinylated nanoparticles produced more fluorescence than the non- biotinylated (amine-functionalized only) nanoparticles, when captured by the streptavidin coated magnetic beads, meaning that biotin-streptavidin binding was taking place on the functionalized surfaces. Detection using dye-loaded nanoparticles was further improved by adding bovine serum albumin (to reduce nonspecific binding) or alginate (to reduce the surface charge of the amines and trap the dye). The SR-101 dye-loaded particles released the fluorescent dye much faster than the FITC labeled particles after the high pH buffer solution was added. Overall, the simple real-time detection assay using fluorescently labeled silica nanoparticles was successful.

638608 Photocrosslinkable Affibodies for Efficient Cancer Treatment and Proliferation Inhibition

Abstract:

One of the major challenges in treating tumorous cancer is delivering an anti-cancer drug deep enough within a solid tumor for a sustained period of time to ensure effective cell killing and to prevent recurrence. Currently, antibodies are often used to target cancer cells due to their affinity for specific overexpressed cell receptors and ability to act as a vehicle to deliver chemo drugs. However, antibodies face several issues in oncology. Due to their large size, antibodies have issues penetrating dense cellular matrices and their tight binding prevents further diffusion inwards from the outer layers of tumors.

To combat these issues, we asked whether an engineered antibody-like protein, called an affibody, can be designed to not only increase permeation in solid tumor matrices but also crosslink a specific cell receptor to impart improved therapeutic efficacy as measured by sustained retention in the tumor bed with increased drug delivery and proliferation of cell growth.

First, recombinant protein engineering was utilized to create an affibody with a benzophenone modification (called N23BP) which provides the ability to covalently photocrosslink with overexpressed epidermal growth factor receptor (EGFR) upon exposure to near UV. This photocrosslinkability was confirmed in solution via SDS-PAGE. Retention in 4T1-cell tumor model spheroids was determined by photoluminescence. Next, N23BP was fused with cytosine deaminase (CodA) to give it the ability to convert prodrug 5-fluorocytosine to the anticancer drug 5-fluorouracil. The time and effectiveness of drug delivery in MDA-MB-468 cells was assessed using confocal microscopy, monitoring ERK activity, and MTT assay. Lastly, the inhibitory effects of N23BP covalently binding EGFR were explored by comparing it to Cetuximab, a commercial antibody which inhibits EGFR.

The research here shows the ability for sustained affibody expression throughout tumor models due the affibody's small size and ability to covalently bind EGFR. Furthermore, the affibody-enzyme N23BP-CodA shows promise of increasing drug delivery time to 48 hours as compared to a wild type non-photocrosslinkable affibody-enzyme or the antibody drug Cetuximab. Recently obtained results also show that photocrosslinking affibodies to cell receptors induces a significant decrease in cell growth and proliferation which could be used in the future as additional modalities for cancer therapy.

638341 Development of pH-Sensitive Hydrogel Nanoparticles for Oral Delivery of High Isoelectric Point Protein Therapeutics

Abstract:

Currently, protein-based therapeutic drugs are only available to patients via injection. This method of drug delivery is uncomfortable to the patient, and thus results in decreased patient compliance. Oral drug delivery of such drugs would be preferable for many patients, but several obstacles in the gastrointestinal tract must be overcome for successful oral protein delivery. Ideally, the protein should be transported across the intestinal epithelium where it can enter the bloodstream. We have developed pH-sensitive polymeric nanoparticle carriers to protect the protein through the acidic upper digestive tract and then swell to release the protein in the higher pH environment of the upper small intestine. For proteins with a low isoelectric point, charge interactions between the protein and the anionic particle facilitate loading and release. However, this is not true for high isoelectric point proteins, as the positively charged protein will remain attracted to the negatively charged particle in neutral conditions, thus hindering release. The present study explores strategies to overcome this attraction between the protein and nanoparticle, and thus, increase bioavailability of the high isoelectric point protein therapeutics.

Hydrogel nanoparticles were synthesized by inverse emulsion polymerization of acrylamide and itaconic acid and crosslinked with N,N'-methylenebisacrylamide. The particles were purified using ethanol precipitation and dialyzed against a gradient of ethanol and water. Particle composition was confirmed with Fourier Transform Infrared Spectroscopy and particle swelling was analyzed with dynamic light scattering to measure particle size at varying pH levels. Electrophoretic light scattering was also used to measure the zeta potential of the particles at varying pH levels. With increasing pH, the particles increase in size, and zeta potential becomes more negative. Protein loading into the particles over time was measured using the Micro BCA protein assay, which showed decreased supernatant protein concentration over time. These pH-sensitive carriers show promise for oral protein delivery. Further studies are in progress to evaluate protein release, cytocompatibility, and transport across the intestinal epithelium.

638392 Effectiveness of Polyelectrolyte Surfactant Nanoparticles in the Eradication of Bacterial Biofilms

Abstract:

Cystic fibrosis (CF), a genetic disease, is characterized by the buildup of thick mucus secretions in the lung. Due to the overproduction of mucus, patients are more susceptible to suffer from chronic bacterial infections, most notably due to *Pseudomonas aeruginosa* in adult CF patients. *P. aeruginosa*, a gram-negative species, exists in complex biofilm communities attached to living or abiotic surfaces. Biofilms are intricate bacterial communities with a protective extracellular matrix that contribute to inflammation and the presence of multi-drug resistant strains. The Roth Lab Group has previously developed a polyelectrolyte surfactant (PS) nanoparticle to aid with the controlled drug release and protection of cationic antimicrobials from degradation. Additionally, PS has the ability to penetrate into mucus and biofilm surfaces due to their amphiphilic nature. Polyelectrolyte surfactants are composed of an anionic poly (methacrylic acid) backbone with 10% Jeffamine M-2070 grafted as pendent chains. The anionic backbone encapsulates cationic antimicrobials via electrostatic self-assembly to form a biocompatible nanoparticle. Current cationic antimicrobials subjected to testing as nanoparticles are tobramycin and polymyxin B. Tobramycin is the gold standard antibiotic used against gram-negative species in cystic fibrosis, while polymyxin B is another antibiotic with activity against gram-negative species. To understand the efficacy of these nanoparticles against bacterial biofilms, their minimum biofilm eradication concentrations (MBEC) were analyzed on clinical isolates of *P. aeruginosa* from CF patients. The MBEC refers to the lowest concentration of an antimicrobial agent required to reduce the biofilm's cell viability from attached surfaces. The results of the clinical strains treated with both nanoparticles and free antimicrobials shows that polyelectrolyte surfactant nanoparticles have consistently lower MBEC values than their cationic antimicrobial counterpart. The analysis indicates the polyelectrolyte surfactant nanoparticles are more efficient in the eradication of biofilms from attached surfaces. Future studies of polyelectrolyte surfactant nanoparticles include testing nanoparticles of various percent grafts and with additional antimicrobials, like polymyxin E.

638471 Sequential Flash Nanoprecipitation for the Scalable Formulation of Nanoparticles Using High Mass Concentration Mixing Streams

Abstract:

In the industrial scale production of pharmaceuticals, maximizing the mass concentrations of mixing streams is of economical benefit as it minimizes the volume scale/throughput required of process systems to generate a given amount of product. Flash NanoPrecipitation (FNP) is a scalable process to formulate nanoparticles with a hydrophobic core and amphiphilic stabilizing shell. Here, we demonstrate a variation of the FNP technique that separates the nucleation and stabilization steps into separate sequential mixers, allowing the hydrophobic active to nucleate and grow unimpeded by adsorption of the stabilizing polymer. Using this sequential Flash NanoPrecipitation (sFNP) technique, we can formulate stable nanoparticles in the 200-400nm range while increasing the mass concentration of the core and stabilizer streams beyond what is possible in a single mixer. This technique provides a process for nanoparticle production in a middle size range that was previously unattainable using traditional FNP. The potential of this scalable platform technology to decouple the size and core loading of particles produced by FNP could provide finer control over formulation development than previously demonstrated for turbulent mixing.

638494 Characterization of Protein Corona on Gold Nanoparticles Via SDS-PAGE

Abstract:

Nanoparticulate drug carriers encounter a vast number of biological compounds when traveling through the human body. The majority of these compounds are proteins commonly found in biological fluids like blood and lung fluids. The proteins in these biological fluids adhere to the exterior of the nanoparticle and mask the particle's surface chemistry. This can be detrimental to therapies that rely on their surface chemistry to illicit a desired response in areas with high protein concentration, like the lower lungs. Zwitterionic polymer coatings can reduce the number of proteins binding to the nanoparticles, thus improving drug potency. In the current study, bare gold and polymer coated gold nanoparticles were introduced to protein-rich human serum and bronchoalveolar lavage fluid (BALF). The proteins that adhered the nanoparticle surfaces were then characterized with 1D SDS-PAGE, to separate the proteins based on size, followed by mass spectrometry, to verify protein composition in the SDS-PAGE bands. We have observed a substantial decrease in the intensity of protein bands between the polymer coating and the raw nanoparticles. The intensity of the bands is directly correlated to the number of proteins binding to the particles' surface. Entire bands that were observed on the bare nanoparticles had disappeared on the polymer coated trials. Future studies will determine the impact the polymer size and coating density on protein adsorption.

638515 Flash Nanoprecipitation of Hafnium Oxide Nanoparticles for Theranostic Applications

Abstract:

Hafnium oxide (hafnia) nanoparticles have demonstrated potential for use in theranostic applications such as computed tomography (CT) and radiotherapy. Thermal composition synthesis methods are of interest for producing hafnia nanoparticles because of their potential to provide better size control and improved cytotoxicity compared to other synthesis techniques. However, thermal composition methods produce particles coated with an organic layer and are hydrophobic. Thus, to be used introduced into the body, it is necessary to prepare the particles using a method that transfers the particles to water while maintaining their stability.

In this study, we investigate flash nanoprecipitation (FNP) as a method of suspending hafnia nanoparticles in water. FNP is a fast, scalable approach to produce composite polymeric nanoparticles through rapid mixing in a multi-inlet vortex mixer. We demonstrate that this technique can be reliably and reproducibly applied to produce composite block copolymer stabilized hafnia nanoparticles in water. Hafnia nanoparticles produced in this work were synthesized through thermal decomposition of hafnia trifluoroacetate in oleylamine. These particles were used to create hafnia nanoparticle clusters (HNCs), featuring a hydrophobic core of hafnia nanoparticles, green fluorophore (3,3′-Dioctadecyloxacarbocyanine perchlorate), and polylactic acid (PLA) stabilized by an outer shell of amphiphilic poly(lactic acid)-b-poly(ethylene glycol) (PLA-b-PEG). These HNCs were evaluated to assess CT performance, stability in cell culture media, and cytotoxicity. From this work we hope to present a novel CT imaging agent with reduced cost and improved biocompatibility to replace current imaging agents.

637580 Design of a Low-Cost Surface-Enhanced Raman Spectroscopy (SERS) Substrate Using a Modified Seed-Mediated Gold Nanoparticle Growth within a Cellulose-Based Filter Membrane

Abstract:

Despite nearly one-third of American women contracting bacterial vaginosis (BV) in their lifetime, there remains no rapid detection platform for the infection. Most detection techniques used today exhibit low specificity, are invasive, or are slow due to the requirement of amplification techniques or overnight cultures. Herein, we describe a foundational protocol for the synthesis of a disposable nanoplasmonic substrate for capturing bacteria using cellulose-based filter membranes. The platform is used in conjunction with surface-enhanced Raman spectroscopy (SERS) to detect the biochemical composition of the captured bacteria. SERS can enhance a molecule's intrinsic Raman signal by 104x or more, negating the need for overnight bacterial culture and more laborious amplification techniques. To be useful for biological detection, the substrate must enhance the Raman signal of target molecules while producing a low background noise from the cellulose paper and other contaminants. Herein, a modified seed-mediated gold nanoparticle growth protocol was established to generate the SERS substrate. Background signal reduction and target molecule signal enhancement were optimized by altering the concentration and size of gold nanoparticles grown throughout the fibers of the cellulose paper. The optimization steps included identifying the ideal cellulose paper soaking duration, reagent concentrations, and drying times to produce a SERS compatible substrate. Reflection measurements using a portable Ocean Optics spectrometer were used to indirectly determine the wavelength range of the plasmonic absorbance. SERS measurements were acquired using a benchtop Raman micro-spectroscopy system at 785 nm excitation. Malachite green isothiocyanate, a common Raman reporter molecule, was used to demonstrate the SERS signal enhancement capabilities of the substrate. The designed substrate was proven to generate a reproducible signal, thus establishing a protocol to create an effective, low-cost substrate optimally suited for small biomolecule detection. In the future, we hope to engineer the selective binding properties of the substrate to be specific to bacteria-associated BV infections.

637775 Molecular Dynamics Simulations Provide Insight into the Structure and Stability of an Aptamer-Target System

Abstract:

Aptamers are single-stranded nucleic acids that behave similarly to antibodies in that they bind to certain target molecules. There are several key differences between aptamers and antibodies that make aptamers interesting molecules to model. The most important difference is that aptamers have tunable properties and are generally more chemically resistant than antibodies.[1] As a result, we can study how the structure and stability of an aptamer can be affected by changing conditions such as pH, temperature, and concentration of aptamer. Studying a specific aptamer in lab requires time and resources. This resource requirement becomes magnified when one wishes to study a library of aptamers. A complementary method for investigation of aptamer-target systems is molecular dynamics simulations. This computational method involves taking the molecule of interest and running a simulation to create a model that gives a prediction of how the molecule will behave in real life. Our chosen aptamer is the malachite green aptamer (MGA) which binds to its target molecule: malachite green (MG). malachite green is an organic molecule mainly used in dyes and fungicides. We focus on using molecular dynamics simulations to generate updated models of how MGA behaves in water in the unbound state and while bound to malachite green. Specifically, it is of interest to determine the mechanism through which MGA binds to malachite green. It has been reported in previous studies that an “induced fit”[2] method is the main way MGA binds to MG. In other words, the structure of MGA is only loosely determined before binding and undergoes a reorganization of structure while it is binding to MG. This raises questions about the overall flexibility and stability changes of MGA after it has bound MG. These changes can lead to a more exposed MG within the binding pocket, leaving it more susceptible to structural attacks. This is significant because the overall behavior of the MGA-MG system can change, such as faster oxidation, color change, and limited or increased fluorescence. In this presentation, we provide an additional insight into the MGA-MG system.

638261 The Impact of Astrocytes on Tumor Cell Dormancy Versus Proliferation in Brain Metastatic Breast Cancer Spheroids

Abstract:

While great strides have been made in the fight against cancer through primary, adjuvant, and palliative treatments, research has shown that cancer cells can stay dormant for extended periods of time, which also allows them to resist treatments. The signals that enable cancer cells to leave the dormant state and proliferate are currently not well understood, particularly in the context of brain metastatic breast cancer. Herein, we examined the impact of astrocytes (supporting cells of the brain) on the phenotype of brain metastatic breast cancer cells in a co-culture spheroid model. In this model, we employed MDA-MB-231Br brain metastatic breast cancer cells and C8-D1A mouse astrocytes in ratios of 1:1, 1:4, and 1:9 with cancer cells only serving as controls and cultured them for a period of 7 days. Among the 3 co-culture ratios tested, 1:4 and 1:1 culture conditions showed an increasing trend in proliferation as measured via spheroid area. The spheroid area for 1:9 condition decreased initially and stayed constant over the period of 7 days. Consistent with this observation, we found that the percentage of Ki67 (a proliferation marker) cells were significantly lower in the 1:9 condition compared to the other conditions. In future studies, we will further characterize the tumor cell phenotype, as well as their response to drug therapies. Such models provide useful tools to study cell-cell interactions and how such interactions impact the dormant versus proliferative phenotype. Ultimately, an improved understanding of dormancy may lead to the development of new therapeutic strategies.

638686 Applying Differential Gene Set Enrichment Analysis to Elucidate Metabolic Vulnerabilities in Tumor Cells Under Oxidative Stress

Abstract:

Applying Differential Gene Set Enrichment Analysis To Elucidate Metabolic Vulnerabilities In Tumor Cells Under Oxidative Stress

Introduction

In MCF-10A human mammary epithelial cells, expression of a constitutively active mutant AKT kinase (mAKT) renders cells sensitive to disruption of glycolysis. Specifically, when mAKT-MCF-10A cells are switched from normal glucose to galactose, mAKT-MCF-10A cells die for ~3 passages (15 days).

Differential Gene Set Enrichment Analysis (DGSEA) is a novel bioinformatic technique developed by the Graham lab that measures how two metabolic pathways are altered relative to each other. Because metabolic adaptations often involve tradeoffs between two different metabolic pathways, DGSEA may lead to novel insights compared to existing GSEA methods. Here, DGSEA can be used to analyze how metabolic pathways are regulated in MCF-10A cells expressing different oncogenes when cells are cultured in oxidative culture.

Methods

We used previously collected metabolomic data from MCF-10A cells expressing red fluorescent protein (negative control) or activated AKT (mAKT) cultured in glucose and short-term galactose for 24h. We ranked intracellular metabolite pool sizes by $\log_2$ fold change between the short-term galactose cells and long-term glucose cells. Next, we used DGSEA to identify pairs of metabolic pathways that are differentially regulated. Large changes in pairs of metabolic pathways were indicated by large, normalized enrichment scores (NES). The p-values and false discovery rates (FDR) were estimated by permutation analysis. The criteria used to determine significant metabolic pathway pairs was a p-value < 0.05 and a false discovery rate < 0.25.

Results and Conclusions

Significant metabolic pathways pairs comparing mAKT-MCF-10A cells placed in short term galactose culture to those placed in glucose culture were analyzed using DGSEA. Four significant pathway pairs were found: pyrimidine and nicotinate/nicotinamide, pyrimidine and butanoate, pyrimidine and glycolysis/gluconeogenesis, and lastly pyrimidine and glutathione. For these pathways, pyrimidine metabolism was the downregulated pathway when mAKT cells were placed in galactose culture. These other paired pathways present potential tradeoff checkpoints that the cell may elect to prefer at the expense of pyrimidine nucleotide metabolism in times of oxidative stress.

Discussion

Our finding that pyrimidine is downregulated when tumor cells are under oxidative stress induced by being placed in galactose culture corroborates findings within the field of metabolomics suggesting that cancer cells rely on increased pyrimidine and nucleotide synthesis pathways to fuel tumor growth. The findings of this project suggest a possible mechanism that oxidative stress induced by the galactose culture led mAKT cells to downregulate pyrimidine synthesis.

638560 Metabolic Pathway Engineering of paclitaxel Biosynthesis in *Taxus Chinensis* Plant Cell Culture

Abstract:

Paclitaxel, a chemotherapeutic produced by the plant *Taxus chinensis*, is one of the most commonly used FDA-approved drugs for treatment of lung, breast, and ovarian cancer and one of few compounds industrially produced using plant cell culture. Commonly used approaches for increasing yield of paclitaxel in plant cell culture, such as methyl jasmonate elicitation, result in decreased cell growth and viability and production of undesired side products, such as phenolics and flavonoids. Thus, the objective of this research is to specifically increase production of paclitaxel via metabolic pathway engineering of the taxane biosynthetic pathway. First, a stable *Agrobacterium* infiltration-based transformation method for *Taxus chinensis* cell lines was established using a pCambia1300 binary vector expressing the GUS+ reporter gene. After optimizing transformation methods, over a 120-fold increase in GUS expression was observed for the *Taxus chinensis* cell line 48.82A.3s one week after *Agrobacterium* infiltration. Stable genomic integration of the selectable marker (hygromycin resistance) 3 months after transformation was also confirmed using a PCR screen on extracted genomic DNA. Next, four taxane biosynthetic pathway genes (BAPT, DBAT, DBTNBT, and TASY) were each overexpressed using the same pCambia1300 binary vector. After transformation, cells were treated with different concentrations of hygromycin (0, 2.5, 5, and 10 mg/L) to determine the minimum inhibitory concentration (MIC). Following two weeks of incubation, cell viability was measured using an Alamar blue assay, expression of taxane biosynthetic pathway genes was measured using RT-qPCR, and paclitaxel content was analyzed using ultra performance liquid chromatography (UPLC). Hygromycin was determined to be a suitable selection agent, with a 97% decrease in cell viability for wild type cells at 10 mg/L. Preliminary results from these experiments indicate that overexpression of DBTNBT resulted in a log2-fold increase in DBTNBT expression of over 16 and a 5.7-fold increase in paclitaxel production when normalized by viable cell count. In order to more successfully identify the transformed cells in these heterogeneous cell populations, we developed a non-destructive method for fluorescence screening of positive clones. A C-terminal mCherry tag was added to the aforementioned taxane biosynthetic pathway genes (along with a 10 amino acid linker) using Gibson assembly and these constructs were then transformed into *Agrobacterium tumefaciens* EH105 using electroporation. This novel non-destructive method will enable us to more easily develop stable homogeneous cell lines and improve our study of taxane biosynthetic pathway engineering.

632261 The Impact of M2-Polarized Macrophages in Hedgehog Signaling Activity and Proliferation of Triple Negative Breast Cancer Cells

Abstract:

Hedgehog (Hh) signaling activity is highly upregulated in about 30-40% of tumor specimens and it has been associated with poor prognosis in breast tumors of the triple-negative sub-type. Hh signaling activity has been linked with the phenotype of tumor-associated macrophages (TAMs) in tumors, yet its oncogenic role within the context of triple negative breast cancer (TNBC) is not well-understood. We hypothesize that TAMs support the oncogenic activity of Hh signaling in TNBC in a paracrine manner. Hh signaling activity was examined in monocytes, TAMs, and MDA-MB-231 (TNBC cell line). Analysis of gene expression levels showed no significant difference in the levels of *GLI1*and *SMO*genes in TAM monocultures +/- SHH ligand, indicating absence of canonical Hh signaling activity. Protein levels of the SHH ligand were only found expressed in TNBC cells. Macrophage cell retention and tumor cell growth was examined in monoculture and a co-culture system at increasing ratios, 2:1 and 1:1 TAM to TNBC ratios. NVP-LDE225 (NVP) was used as a pharmacological suppressor of the Hh pathway. Macrophage retention in the co-cultures were analyzed in comparison to the TAM monocultures, these demonstrated no statistical difference across all conditions. Initial examination of growth rates indicates that NVP does not affect the TAM and TNBC monocultures; having a non-significant difference when analyzed through a One-Way ANOVA. For both co-culture ratios, Tukey's multiple comparisons test indicated that the TAM promotes tumor cell proliferation, and this effect was independent of NVP treatment indicating that the Hh pathway is not involved in this paracrine interaction. Future work will seek to examine this paracrine interaction in the presence of fibroblasts which are a well-known cell target of Hh inhibitors. This will allow to determine the impact of TAM in the oncogenic activity of the Hh pathway within the context of the tumor-stroma interaction.

638422 A Cell-Free Pipeline for Rapidly Imprinting Complex Methylation Patterns on DNA to Enhance Plasmid Transformation in Bacteria

Abstract:

There has been an increasing interest in the development of bacteria for use as next-generation industrial biocatalysts or as cell-based diagnostics and therapeutics. Bacterial genotype-phenotype relationships and microbe-host interactions are poorly characterized, partially due to formidable and strain-specific barriers to DNA transformation. One of the most common approaches to overcoming these barriers is to overcome restriction-modification (R-M) systems by mimicking the DNA methylation pattern which confers self-nonself recognition. However, this approach is limited by the required expression of a host's numerous DNA methyltransferases by the cellular system. This occurrence often results in methyltransferase-induced toxicity and incomplete methylation, especially for complex Type I systems. Here we describe an alternative approach called IMPRINT, which involves the use of cell-free transcription-translation systems to co-express several methyltransferases. IMPRINT rapidly recreates the host's DNA methylation pattern *in vitro* thereby boosting DNA transformation efficiencies. We demonstrate that IMPRINT can be used to readily reproduce DNA methylation patterns present in various hosts and we further optimized this approach to achieve complete multiplexed methylation when complex Type I R-M systems are present. We show that IMPRINT can achieve the same maximum transformation efficiency in strains with known R-M barriers and high transformation efficiencies in recalcitrant strains of probiotic Bifidobacteria. Finally, we demonstrate how barcoding shuttle vectors can rapidly reveal the minimal methylome to achieve maximum DNA transformation in a given strain when more than two R-M systems are present. IMPRINT offers a rapid means of enhancing DNA transformation across the bacterial world.

638237 MYC Pathway Inhibition Leads to Dormancy in Leukemia Cells

Abstract:

Each year, the number of new cases for leukemia in children increases by 7%, equaling around 5 in every 100,000 children. Even though the number of deaths is decreasing each year by 2.8%, leukemia is still a major concern in children at is equals 24.3% of all new cancer cases. Survivability is measured by using the five-year survival percentage that measures the percent of individuals in remission surviving for five years. Survival rate for children aged 15 and below is 68% with 15+ being 66%. Even though the survival rate is quite high, 15 – 20% of leukemia returns which is equivalent to 600 children relapsing. The survival rate after the first relapse drops dramatically to 30-50%. A solution needs to be found in order to stop relapsing from happening.

Cancer cells are some of the fastest proliferating cells that exist and is the reason for chemotherapy drugs to target faster proliferating cells. However, with this comes side effects that lead to other faster proliferating cells to be attacked as well. With chemotherapy, most cancer cells are attacked, and most people are pushed forward into remission, apart from a small fraction of cancer cells remaining. These cells survive the initial attack from chemotherapy drugs as they enter a quiescent state giving them protection from the chemotherapy drugs. This quiescent state is a reversable state in which a cell does not divide but can begin cell division again.

The goal of this research is to investigate a possible reason for these cancer cells to enter this quiescent state and understand what specific function is used to push cells into this state. The focus was on the inhibition of the MYC gene pathway which is used as a regulator gene and protooncogene that are used to encode transcription factors.

637117 Tuning Expression Output Using Non-Native Sigma Factors and Variable Length Spacers

Abstract:

Among the primary goals of synthetic biology is to establish a greater control of genetic systems. To be successful, genetic parts must offer predictable and robust control of output expression that is also orthogonal to native metabolic networks. Non-native sigma factors, which initiate the transcription process through interaction of the RNA apoenzyme with their cognate promoter, are an attractive method of control; however, it is still unclear how the recognition of sigma factor binding elements influences transcriptional output, specifically the spacer length between sigma factor binding sites. Our work characterizes the expression of a library of nine synthetic promoters in response to three non-native sigma factors. The designed promoters were derived from the wild type general stress response protein genes from *B. subtilis* and contained a variable spacer length between 10 to 18 base pairs. By altering the spacer length, we observed a 46-fold variation in expression output for sigma factor B, with the highest fluorescence observed at 13-14 base pairs. We next evaluated the orthogonality of our promoters by testing them with sigma factors F and G from *B. subtilis*, finding similar distributions with a 38- and 52-fold change respectively. We are currently testing libraries of spacers that naturally respond to sigma factors F and G to assess the universality of our hypothesis. Ultimately, our findings indicate that spacer length is an important factor when tuning expression to desired levels, providing another control point for synthetic biologists.

636657 Screening of Norbaeocystin Methyltransferase Variants Enables Enhanced Psilocybin and Baeocystin Production in *E. coli*

Abstract:

While psilocybin is currently under clinical evaluation for the treatment of severe depression, post-traumatic stress disorder (PTSD), and addiction, anecdotal evidence from recreational users has led some to postulate that the ratio of naturally occurring psychoactive metabolites in various mushroom species may greatly impact the experience and overall effect on the brain. This “Entourage Effect” may necessitate a specific production strain capable of high baeocystin production to meet necessary pharmaceutical demand. Norbaeocystin methyltransferase (psiM) is responsible for the iterative methylation of norbaeocystin, resulting in the production of psilocybin via a monomethylated intermediate, baeocystin. The genetic variation between mushroom species and resulting differences in their metabolite profile, have motivated this study to evaluate a range of psiM variants for their ability to selectively produce the intermediate product baeocystin. Here, we examine the effect of genetic variability of the psiM gene and attempt to create a biosynthetic production platform for producing high titers of baeocystin and higher titers of psilocybin than previously reported. Screening of genes from *Psilocybe cubensis*, *Psilocybe cyanescens*, *Panaeolus cyanescens*, *Gymnopilus dilepis* and *Gymnopilus junonius* in a high throughput well plate assay yielded a production strain capable of producing more than 480 mg/L of psilocybin prior to fermentation optimization and scale up. This surpasses the current best performing psilocybin production strain by an increase of 245% under similar conditions. Additionally, initial screening produced a range of strains with enhanced baeocystin production. Ongoing efforts to validate, optimize, and scaleup multiple lead strains is currently underway. This work further enhances the viability of industrial batch production of psilocybin using fermentation and provides and starting point for the development of strains capable of selectively producing its precursors for combined medicinal use.

637743 Inducible Directed Evolution (IDE) of Complex Phenotypes in Bacteria

Abstract:

Many important microbial phenotypes are complex and difficult to engineer, as they emerge from interactions between many genes. Directed evolution is a method of engineering beneficial phenotypes in microbial organisms via an iterative process of mutagenesis and screening, improving function without the need for detailed biochemical models. Directed evolution often suggests novel or counterintuitive routes to improved function, making it ideal for engineering complex phenotypes and giving it diverse applications in natural product synthesis and microbial therapeutics. Recent methods for directed evolution have overcome the traditionally laborious and costly steps for generating error-prone libraries, which include molecular cloning and transformation. However, all methods for directed evolution developed to date are limited to small regions of DNA and/or allow the accumulation of off-target mutations. To enable directed evolution of complex phenotypes encoded by multigene pathways, we require large library sizes for DNA sequences >5-10 kb in length, elimination of off-target genomic mutations, and decoupling of diversification and screening steps.

To meet these challenges, we developed Inducible Directed Evolution (IDE), which uses a temperate bacteriophage to inducibly package large plasmids and transfer them to naive cells after intracellular mutagenesis. IDE combines (1) the P1 bacteriophage's ability to deliver large plasmids efficiently to *Escherichia coli* cells and (2) inducible intracellular mutation machinery to introduce and propagate random mutations in the target pathway without the accumulation of off-target mutations. We demonstrate no reduction in library size with increasing pathway length, up to the largest pathway we have tested (35 kb). We also demonstrate IDE's utility in evolving several complex multi-gene pathways, including a 5-gene tagatose catabolism pathway (5 kb) from *Bacillus licheniformis*, as well as a 10-gene melezitose catabolism pathway (15.4 kb) from *Bifidobacterium breve*.

638240 Electrochemical and Electrogenerated Chemiluminescence of Tris(2,2′-bipyridyl)Ruthenium(II)-Based ITO Electrodes

Abstract:

Electrochemiluminescence (ECL) is an analytical technique which molecules undergo an electrochemical reaction on an electrode surface when a certain voltage applied that allows energy transfer of electrons to occur. The light emitted can then be quantified to observe the number of molecules which underwent the ECL process therefore is applied, the ECL method provides numerous advantages as a biosensing analysis method for disease biomarkers. Our objective was to further develop the ECL analysis via functionalization of ITO electrodes allowing us to observe the chemiluminescence reaction to occur via the /TPA system. Therefore, to accomplish this, the following action items were conducted: Develop two different functionalization protocols of ITO electrodes and analyze ITO electrodes functionalization via Electrochemical Impedance Spectroscopy (EIS) to determine the most effective method toward observing photon detection with charge resistance transfer levels for ECL. Determine redox peak via cyclic, linear sweep, and square wave voltammetry to evaluate the most effective voltage range for ECL stimulation in functionalized electrodes. The redox peak allowed us to observe where a redox reaction occurs with a voltage applied is most evident from 1.2-1.9V. Utilize ECL analysis software with ITO electrodes and observe if electrochemical reaction occurs and statistically significant photon emission is observed. One of the surface functionalization approach (ITO electrode/APTES/BS^3/ Ru(bpy)-streptavidin) was the most effective method for observation of photon emission due to the lower transfer resistance observed which would allow for more surface interaction to occur during the ECL experiment.

638276 Cohort Analysis Reveals Subtype-Specific Effects of Obesity on the Colorectal Tumor Transcriptome

Abstract:

In this study, we investigated the consensus molecular subtype (CMS)-dependent effects of obesity on the colorectal cancer (CRC) tumor transcriptome. Colorectal cancer is the third most common cancer and the third leading cause of cancer-related deaths in the U.S. Obesity, a worldwide public health concern, is a known risk factor for cancer including CRC. However, the mechanisms underlying the link between CRC and obesity have not been fully elucidated in part because of the molecular heterogeneity of CRC. In 2015, an international consortium classified CRC into four CMSs with distinct characteristics. Yet, it is not known whether the mechanisms linking obesity and colon cancer are molecular subtype specific. Thus, we examined the transcriptomic profile of tumors from obese patients compared to healthy BMI patients. RNA-sequencing data and associated patient body mass index (BMI) data for 231 patients were obtained from The Cancer Genomic Atlas - Colon Adenocarcinoma (TCGA-COAD) Program. Samples were classified into the four CMSs with the CMScaller R package, while calculated BMI was used to categorize patients as normal, overweight, or obese. Gene Set Enrichment Analysis (GSEA) was conducted to compare obese and normal categories within each CMS. Differentially expressed genes (DEGs) were uncovered with the DESeq2 R package.

We observed that cancer-associated inflammatory pathways such as the IL6 JAK/STAT3 and interferon gamma response were enriched in obese samples compared to normal samples in CMS 1, 2, and 4, indicating that obesity can upregulate similar oncogenic pathways in CRC patients with different subtypes. We also found that inflammatory M1 macrophages and cancer-associated fibroblasts were enriched in obese samples compared to normal samples in the immune infiltrated subtype CMS1 and the stromal infiltrated subtype CMS4, respectively, suggesting that obesity can enhance distinct CMS features.

Moreover, when obese and normal patients were compared for each CMS, the obtained DEGs differed between the subtypes, enabling the formation of CMS-specific obesity-associated gene signatures. Further analysis revealed that high expression of these gene signatures was associated with lower patient survival in CRC cohorts.

Overall, our findings suggest that obesity can enhance known oncogenic inflammatory pathways in CRC and has CMS-dependent effects on CRC. Obesity-associated gene signatures were also identified for each CMS and may have prognostic value.

638618 *Chlorella vulgaris* growth from Sodium Bicarbonate Produced Via Carbon Dioxide Absorption Using Sodium Hydroxide

Abstract:

Chlorella vulgaris is a versatile species of algae which can be used for many different applications, including cabin air cleaning for a NASA space camp on Mars, which is of particular interest. For a space camp application in Mars, whose atmosphere contains 95% carbon dioxide, it can be advantageous to grow algae from an easily optimized aqueous carbon solution rather than from the ambient atmosphere. Additionally, by absorbing cabin air (or atmospheric) CO_2 using Sodium Hydroxide, a more efficient removal of gaseous carbon can be achieved than from removal directly using algae. To determine the viability of *C. vulgaris* growth from inorganic carbon sources in solution rather than from gaseous carbon dioxide, a test was conducted comparing algae growth from 1.9 g/L bicarbonate solution and that from bubbling ambient air (0.07% CO_2) at 1 L/min. The bicarbonate feed solution was prepared by bubbling CO_2-rich air through a sodium hydroxide solution using a bubble column. Considering the effluent CO_2 fraction versus time, the CO_2 effluent in the exiting gas stream was initially very low, indicating a high absorption of CO_2 into the solution. Then, the effluent gas composition rose to approximately one half of the inlet and was stable for a time. By the time that this effluent fraction began to stabilize, the vast majority of the hydroxide ions were consumed to form CO_3^{2-}. It is during the subsequent stabilization period that the CO_3^{2-} ions transitioned into HCO_3^- with the addition of more CO_2 gas. Comparing the growth rates of the two algae trials, the algae in the bicarbonate reactor grew more quickly during days 1-7 but slowed behind the ambient air reactor during subsequent days. The bicarbonate reactor peaked during days 5-8 containing 0.6 g/L algae, while the ambient air reactor reached 0.8 g/L algae at day 11. The pH of the bicarbonate-fed algae solution rose to over 11 by day 11, which was a possible cause of slowed growth. Overall, this trial indicates the promise of bicarbonate-fed *Chlorella vulgaris* for use in this space-camp application and further expands the possibility of a human colony on Mars.

638543 Effects of Methanobactin on the Degradation of Methylmercury

Abstract:

Exposure to even small amounts of methylmercury can cause serious health problems, and is especially damaging to unborn infants as their brains and nervous systems are developing. In natural environments certain anaerobic microorganisms are the predominant cause for the production of methylmercury (MeHg) toxin. Methanobactin (MB) is a chalkophore (copper binding) secreted by some methanotrophs. The ligand set responsible for copper-binding can also bind different mercury species. The best-characterized MBs to date are those from Methylosinus trichosporium OB3b and Methylocystis strain SB2. Previous studies have indicated methanotrophs that produce and take up MB are able to degrade MeHg. We wanted to see if the presence of MB on methanotrophs that don't produce MB such as, Mm. album BG8 wild type and ΔmbnT, Methylocystis sp. strain Rockwell, and Methylococcus capsulatus Bath, would affect MeHg degeneration. In order for MB to uptake methylmercury, a TonB-dependent transporter. Mm. album BG8 and Methylocystis sp. strain Rockwell were found to have genes similar to mbnT that encode the transporter. Experiments were performed in the absence/presence of MB from either Ms. trichosporium OB3b or Methylocystis sp. strain SB2. Our data provided evidence that some methanotrophs "steal" MB and such "theft" enhances the strains' ability to degrade methylmercury. Both Mm. album BG8 and Methylocystis sp. strain Rockwell had increased MeHg degradation when treated with OB3b or SB2 MB than in the absence of MB. However, this was not the case for all the methanotrophs. MeHg degradation observed in Mc. capsulatus Bath, and ΔmbnT mutant of Mm. album BG8 were not significantly different in the presence or absence of either form of MB. There are many diseases associated with excess metal in the body. I'm sure the first ones that come to mind are lead and mercury poisoning, but there's many more. There's currently research being done for Wilson's disease which is a rare genetic disorder characterized by excessive copper in various body tissues. Through better understanding of methanotrophic communities, we can bring the world one step closer to finding treatment for just methylmercury poisoning and other medical issues in situ.

636764 Homebrewed Psilocybin: Can New Routes for Pharmaceutical Psilocybin Production Enable Recreational Use?

Abstract:

Psilocybin, a drug most commonly recognized as a recreational psychedelic, is quickly gaining attention as a promising therapy for an expanding range of neurological conditions, including depression, anxiety, and addiction. This growing interest has led to many recent advancements in psilocybin synthesis strategies, including multiple *in vivo* fermentation-based approaches catalyzed by recombinant microorganisms. In this work, we show that psilocybin can be produced in biologically relevant quantities using a recombinant *E. coli* strain in a homebrew style environment. In less than 2 days, we successfully produced approximately 300 mg/L of psilocybin under simple conditions with easily sourced equipment and supplies. This finding raises the question of how this new technology should be regulated as to not facilitate clandestine biosynthesis efforts, while still enabling advancements in psilocybin synthesis technology for pharmaceutical applications. Here, we present our homebrew results, and suggestions on how to address the regulatory concerns accompanying this new technology.

636903 Tissue-Engineered 3D Trabecular Bone Model to Recapitulate Bone Remodeling

Abstract:

Trabecular bone is characterized by a 3D sponge-like porous structure, which undergoes bone remodeling in response to mechanical and chemical stimuli. Reproducing this process in vitro is imperative to understand trabecular bone metabolism in health and disease. Yet it remains a difficult task due to the lack of experimental models that can recapitulate key 3D structural and mechanical aspects of trabecular bone. We aimed to develop a model that can overcome this need. Here, we introduce a 3D trabecular bone tissue model using demineralized trabecular bone (DTB) scaffolds and a cyclic compressive bioreactor in a multi-well plate to induce mechanical stimulation. Our model represents an opportunity to provide new insight into how bone remodeling homeostasis is maintained and its role in diseases such as bone cancer and osteoporosis.

DTBs were prepared by demineralizing bone in 1.2N HCl solution. Osteoblasts were isolated from DsRed mice and seeded onto the DTBs. To induce chemical stimulation, vitamin D3 was used. To induce mechanical stimulation, we developed a multi-well plate-based mechano-culture platform consisting of three parts; (i) a pneumatic actuator moving a magnet array up-and-down, (ii) a magnetic spring-based insert to compress the DTB, and (iii) a microprocessor to control actuator movement.

Our DTB provides advantages, such as optical transparency with high porosity and surface area. Dsred osteoblasts adhered and proliferated well, coating the surface of the DTB within one week. In our platform, a pneumatic actuator receives air and pushes a magnet array closer to the well-plate, attracting a magnetic insert down, compressing the DTB. As air is removed, the array lowers, releasing the DTB. This is repeated in 1 min on/off cycles. In our proof-of-concept studies, osteoblasts were responsive to both chemical and mechanical stimulation, which was confirmed by the alteration of OPG and RANKL secretion profiles. Our ongoing study investigates four conditions to substantiate the relative impact of different stimuli; (1) mechanical, (2) chemical, (3) mechanical and chemical, and (4) no stimulation.

We have developed a 3D trabecular bone tissue model which incorporates both the mechanical and chemical aspects of trabecular bone remodeling to provide a reproducible and anatomically relevant model. Soon, bone marrow monocytes will be included to better simulate the bone remodeling cellular process. We envision a 3D dynamic bone tissue model that can provide better insight into bone metabolism and facilitate biomedical research into various bone remodeling associated diseases including bone metastasis and osteoporosis.

637883 *Staphylococcus Epidermidis* Growth in Mammalian Cell Culture Media for Models of the Host-Pathogen Interface

Abstract:

Staphylococcus epidermidis Growth in Mammalian Cell Culture Media for Models of the Host-Pathogen Interface

Ciara R. Young[1], Lily I. Gaudreau[1], Elizabeth J. Stewart, Ph.D. [1,2],

1. Department of Chemical Engineering, 2. Department of Biomedical Engineering, Worcester Polytechnic Institute, Worcester, MA

Staphylococcus epidermidis biofilms frequently cause hospital-originating infections and sepsis. Clinical biofilms form in the host environment, which can alter biofilm behaviors. Models of the host-biofilm interface are required to study how the host interface influences biofilm growth and antimicrobial tolerance. An important step towards the creation of an *in vitro* model of the host-biofilm interface is establishing and assessing biofilm growth in mammalian cell culture media. Here, we report on the effect of mammalian cell culture media and initial bacterial cell concentrations on the planktonic and biofilm growth of *S. epidermidis.* We found that initial cell concentrations did not alter planktonic growth rates when comparing growth in mammalian cell culture media to growth in bacterial culture media. Biofilm growth was assessed though image analysis of confocal laser scanning microscopy images. We report that higher initial cell concentrations are required for achieving biofilm growth in mammalian cell culture media with similar structures to biofilms grown in bacterial culture media. This shows that the biofilm growth and development seen in standard bacterial culture conditions can be recapitulated in mammalian cell culture media when higher initial bacterial cell concentrations are used to grow biofilms. This finding is important for developing *in vitro* models of the host-biofilm interface as mammalian cell culture media is required for maintaining mammalian cells in these experimental models.

638068 Comparative Study of Enzymatic Inhibition in pH Buffering Barriers in Ostomy Care

Abstract:

Over 140,000 ostomy surgeries are performed in the United States annually. This procedure involves creating an artificial passage between the lumen of a hollow organ and the skin to pass waste material (dejecta) and is performed to treat conditions including autoimmune or inflammatory bowel disease, trauma, and cancer. These patients add to the existing population of 750,000-1 million Americans living with ostomies, reliant on external collection products to replace the dejecta retention functionality of the excised or bypassed colon, small bowel, or bladder. Bodily waste is excreted out of the surgical opening known as the stoma and collected into an ostomy pouch. The pouch is affixed to the patient's abdomen around the stoma via a skin-protecting hydrocolloid barrier that is designed to absorb and manage moisture while maintaining load-bearing adhesion.

Despite these design features, leaked dejecta from the stoma can come in contact with the skin surrounding the stoma, leading to skin issues such as moisture associated skin damage (MASD), peristomal contact dermatitis, and enzymatic injury. Ileostomy dejecta typically has a neutral pH and is rich in enzymes that retain pH dependent activity. These neutral pH environments create optimal conditions for high enzymatic activity that can lead to irritation and breakdown of the skin, which normally has an acidic pH.

The relationship between pH level and enzymatic activity in ostomy care is an active area of research. To better understand this relationship, pancreatic proteases commonly found in dejecta (trypsin, chymotrypsin, elastase) were studied *in vitro* in different buffered pH systems. Next, marketed products commonly used as ostomy skin barriers were also studied to understand their ability to buffer pH and the resultant effect on enzymatic activity. These enzymes were reconstituted in the various buffered pH solutions, and their protease activity was measured kinetically using a fluorescence-based assay.

Enzymatic activity was shown to have a strong pH dependence both in samples of individual enzymes and of various combinations. The activity of commonly found enzymes as a function of pH was determined, and this information can help elucidate their role in peristomal skin complications in ostomates. Furthermore, some marketed products were shown to be more effective than others in buffering to an acidic pH leading to reduced enzyme activity. These results provide support for the ability of certain marketed ostomy barrier products to reduce enzymatic activity on peristomal skin through pH modulation.

637730 Classifying Cardiac Fibroblasts with Predictive Modeling

Abstract:

Cardiac Fibroblasts (CF) are a type of connective cell in the heart that is known to play an integral role in cardiac fibrosis, a type of heart disease, following injury. Activated CFs are correlated with the onset and progression of disease. Recent cell studies have shown that various in vitro conditions lead to the activation of CFs. Traditionally, activated CFs are identified by visually inspecting the α-smooth muscle actin (aSMA) macrostructure through immunostaining. A more fibrous aSMA structure suggests the cells are in the activated state, while the lack of fibers suggests a non-activated state. In addition to displaying aSMA fibers, activated CFs have been correlated with larger cell sizes. A more quantitative analysis of these morphological trends is necessary to better understand the characteristics of CF cells and develop a model that classifies activated vs. non-activated CFs from a single cell population.

This poster presents a quantitative analysis of various morphological properties for CF cells in both the activated and non-activated states. Additionally, it demonstrates predictive modeling methods to automatically determine the activation state of CFs. Using fluorescent microscopy combined with computer vision techniques, we computed 15 unique morphological factors of the cells. These properties were fed into the various predictive modeling methods, with the best model reaching 91.2% accuracy.

Using the calculated morphological features, we demonstrate previously described unquantified trends in the literature. We have quantitatively determined that (a) activated and non-activated CF cells contain two unique populations with statistically different morphologies, and (b) activated cells contain a larger colocalization between the aSMA and actin structures compared to non-activated cells. With our computer vision predictive modeling methods, we significantly decrease the classification time of CF cells while minimizing scientist bias.

638162 Platelet Aggregation Study in Arterial Cavities Using the Lattice-Boltzmann Algorithm

Abstract:

Hemostasis is the mechanism that leads to the cessation of bleeding from a blood vessel affected by accidents. Initially, it uses multiple interlinked steps to build a platelet-made plug that closes up the injured site in a process known as platelet aggregation. Despite being widely studied, this process is not yet fully understood, given the complex and tightly intricated relations between the transport phenomena and the biological processes happening simultaneously around the wounded areas. Therefore, many in silico studies have been used to model different stages of hemostasis and have proven useful to solve those complex relations using traditional numerical methods. However, they all share the downside of requiring high computational resources. Hence the need to develop a method suitable for computational parallelization to decrease the simulation times. The medical community could use such computational models to develop patient-specific therapies once the processing times require less than a few minutes to simulate complex scenarios. For those reasons, this work focuses on developing an algorithm based on the Lattice-Boltzmann method (LBM) to reproduce the platelet aggregation process. LBM is an alternative to the traditional numeric methods capable of reducing computing times. The developed algorithm models a cylindric blood vessel as a 2-dimensional rectangle in which a wound is imposed on the lower wall. Once the vessel's wall is damaged, a series of reactions take place around the wounded area attracting platelets to build a plug. Different simulations were carried out to assess the effect of different flow and chemical conditions over 1) The shape of the plug and 2) The platelets concentration inside the plug. In general, the obtained results show good agreement with previous works. More importantly, the results suggest that the shear rate strongly affects the concentration profiles of the platelet-activating species. It was found that there may be a positive feedback loop in which the shear rate initially activates some platelets (due to surface proteins unfolding) that in turn release activating chemicals that are dragged and diffused towards specific areas where platelet activation is enhanced even more. This conclusion could be an important breakthrough because most of the state-of-the-art research on the shear rate influence over platelet activation focuses on effects at a molecular scale, but the current approach is based on a systems biology approach. However, the computational algorithm still requires further refinement to reduce the computation times.

637812 Transaminase Screening for Biocatalytic Functionalization of Polymer Plastic Waste

Abstract:

Exploring enzymatic routes in order to generate high value monomers or polymers offers an opportunity for recovery and diversification of plastic waste. Transaminases (TAs) are a class of enzymes that reversibly convert an aldehyde or ketone to an amine. These enzymes are well categorized for pharmaceutical applications, and this work aims to introduce these enzymes into pathways for plastic valorization centered around terephthalic acid (TPA) derived from depolymerized polyethylene terephthalate (PET). In this project, thirty transaminases were selected for screening and twenty-three were successfully expressed in the *E. coli* ROAR strain. Qualitative colorimetric assays along with quantitative high performance liquid chromatography (HPLC) analysis were used to track TA activity on various substrates. Two library members of previously uncharacterized TAs showed activity on a model substrate, benzaldehyde, and will be the basis of subsequent testing and comparison to previously characterized TAs. Substrate evaluation involved testing three derivatives of terephthalate (TPA), with IPA as the amine donor. Six total TAs were tested, and it was shown that *Chromobacterium violaceum* (C.v) and *Vibrio fluvialis* (V.f), had activity on plastic derivatives, paving the way for future plastic functionalization work. Continued bioprospecting involving this library will allow for the discovery of substrate specificity, thermal tolerance, and solvent tolerance for TAs with potential to create diamine monomers for high value, recyclable plastics.

638610 Quantifying the Driving Ability of Wheelchair Users

Abstract:

People who use electric powered wheelchairs require driving skills training to navigate safely and independent in the community. However, the current methods for evaluating the driving abilities of wheelchair users are subjective, binary (pass/ fail) and may depend on the therapist's experience. Creating a benchmark scoring tool based on a number of quantitative metrics and level of wheelchair driving skills would aid in identifying methods to quantify a wheelchair user's driving ability. This will enable health care providers to monitor the client's mobility progress and personalize driving training to better suit their client's growing and changing mobility needs. Ten wheelchair users were recruited to compare their driving skills in a common driving task. 'Expert' participants were 54.6 +/- 11.7 years of age with over 3 years of wheelchair driving experience. 'Novice' participants were 54.6 +/- 13.3 years of age with over 3 months of driving experience. Five 'experts' and five 'novices' were asked to drive backwards for three trials. A datalogger placed at the center of their wheelchairs recorded linear accelerations and angular velocities in three axes. The variability of each variable was measured using the sum squared error function (SSE). The results showed a significant lower lateral linear acceleration ($p<0.05$) in expert wheelchair users (34.40 +/- 14.42) compared to novice wheelchair users (85.49 +/- 71.18). Similar significant results ($p<0.05$) were obtained in the z-angular velocity (own axis) for experts (0.21 +/- 0.14 e+05) compared to novice wheelchair users (1.99 +/- 2.52 e+05). In order generate a benchmark scoring system, more participants are needed to implement an accurate baseline demonstrating experience with wheelchair ambulation. Numerous subsequent studies with other quantitative metrics and level of impairment must be carried out to take into consideration the progression of degenerative diseases and their influence on fine and gross motor control.

638367 Hydrophobic Drug Loading into Thermosensitive Hydrogels Impacts Rheological Properties

Abstract:

Severe burn wounds lead to a weakened immune system and increased risk of exposure to various common microbes found in hospitals, leading to severe infection. Current treatment includes spreading an antimicrobial ointment or cream on the wound, which can be painful. We have been developing sprayable, antimicrobial, polymer treatments that can be applied to an infected wound as a cold liquid but will transition to a gel state when in contact with the warmer tissue. This thermosensitive hydrogel has the advantages of being applied without pressure, reducing application pain, in addition to providing a cooling effect and the ability to prolong drug release.

Differences in final gel properties were noted in our prior work when the drug was added at different times during the formulation process. Therefore, we aimed to understand how these changes in drug addition timing impacted gel viscosity, viscoelastic behavior, transition temperature, and gel stability in this work. Three methods of addition of a hydrophobic drug (diclofenac salt) were tested: 1) preparation of the bare gel (Pluronic F127, a triblock copolymer with a hydrophobic central block, in water) followed by addition of diclofenac the following day, 2) combining the Pluronic F127 and diclofenac in water followed by an overnight resting period, and 3) preparing the bare gel using ¾ of the water and a drug slurry using the remaining ¼ of the water, followed by the combination of the two mixtures the following day.

These methods were tested for Pluronic percentages of 15, 17, and 20 w/w%; and diclofenac salt percentages of 0.5, 1.0, 1.5, and 2.0 w/v%. Bare gels were also tested for methods one and three to compare gel strength with and without later hydration. Gelation of the Pluronic's occurs due to micellization of the polymers followed by micelle packing when heated. Depending on when the drug is added during the gelation process, it can enter into or be excluded from the micelles. The three methods tested the addition of diclofenac salt once the micelles have been completely formed, as they are forming, and after formation but with additional hydration. Preliminary results show a minor difference in gelation temperature, up to 2.0 °C, but more significant differences in storage modulus or gel rigidity between methods. Future work will include micelle imaging, sizing, and drug release studies.

636694 Development and Application of T7 Promoter Libraries to Enable Psilocybin Production in *Vibrio Natriegens*

Abstract:

Title: Development and Application of T7 Promoters Libraries to enable Psilocybin Production in *Vibrio natriegens*

Author List:

Claire M. Cashdollar

John D. Brinton

Alexandra M. Adams

J. Andrew Jones

Psilocybin is a psychedelic prodrug that occurs naturally in "magic mushrooms", which include over 200 species of fungi, most notably the *Psilocybe cubensis*. Psilocybin is becoming increasingly important in the medical field, as this drug has been shown to be an effective treatment for a multitude of mental disorders, as well as addiction. Previously, commercial production of psilocybin has been achieved through heterologous gene expression in the highly engineerable *E. coli.*

Vibrio natriegens is a gram-negative bacterium that has recently been highlighted as a desirable host for protein synthesis; they have the fastest growth rate known for a non-pathogenic microbe. There are many potential applications for *V. natriegens* in bioengineering, including as a host for the for the recombinant production of psilocybin.

Here, we study the gene expression of the psiD, psiK, and psiM genes, with the purpose of producing psilocybin and its intermediates through a batch-fed *in vivo* approach utilizing a 4-hydroxyindole starting substrate. Because of the rapid-growing nature of *V. natriegens,* these organisms are very appealing for commercial use. Their fast growth enables robust processes with enhanced productivity leading to promising industrial processes with short residence times and limited opportunity for contamination due to the rapid growth.

Here, we implement the first instance of a T7 promoter optimization strategy in *Vibrio natriegens*. Promoter libraries were constructed for *V. natriegens* using the ePathOptimize approach and evaluated in *E. coli* and *V. natriegens*. Interestingly, we found that although the native T7 sequence demonstrated high strength in both hosts, the reduced strength promoters showed a large degree of host specificity. Three T7-variant promoters (T7 consensus, E2, and F2) were developed and used to construct three promoter libraries in operon, pseudooperon, and monocistronic forms. The psiD, psiK, and psiM genes were cloned into each of these promoter library configurations and were screened for norbaeocystin and psilocybin. Efforts are ongoing to develop enhanced production methods for psilocybin in *V. natriegens*, however, preliminary results and comparison to an *E. coli* host indicate that *V. natriegens* may not be the most suitable host for this pathway.

638504 The Threefold Polymorph of Amyloid Beta Fibrils Employ a Distinct Intermediate Frustrated State for Growth

Abstract:

Alzheimer's disease ranks as the third leading cause of death amongst adults over the age of 60 behind heart disease and cancer. The accumulation of amyloid plaque and neurofibrillary tangles causes the irreversible degradation and destruction of neurons that drastically shrinks brain tissue. Amyloid plaques are caused by the buildup of insoluble amyloid beta (Aβ) fibrils formed from self-assembled propagation of peptides. Studies have shown that Aβ can take shape as soluble oligomers or insoluble fibrils with different morphologies. Aβ fibrils attain numerous structures, called polymorphs, both in patients' brains and *in vitro*. Prior observations of fibrils grown *in vitro*, revealed that twofold polymorph of Aβ40 fibrils was dominantly expressed in agitated solutions; whereas the threefold polymorph of Aβ40 fibrils preferentially formed in the quiescent solutions, which also closely resemble the structures of Aβ fibrils extracted from Alzheimer's patients' brains.

Previous atomic force microscopy (AFM) studies have demonstrated that the twofold symmetric Aβ40 polymorph grew steadily at broad ranges of peptide solution concentrations. It was also established that the fibril growth rate was determined by an activated complex, comprised by a peptide chain partially bound to the fibril tip and folded in a conformation that is distinct from that of the fibril bulk; the rearrangement of the intermediate configuration of the incoming peptide to the structure in the bulk fibril constituted the transition state for growth [1]. Using time-resolved *in situ* AFM, the growth of seeds nucleated in quiescent solutions were monitored, and the data showed that the correlation between the growth rate with the peptide concentration in the solutions drastically diverged from the linear kinetic dependence shown by the twofold fibrils. For the threefold polymorph, the growth rate increased linearly with peptide concentration up to 5 mM, but at higher concentration it reached a saturation value. A model, analogous to the Michaelis-Menten law for enzyme kinetics, was developed to describe the kinetics of the frustrated complex at the fibril tip. Further analyses of the model have suggested that the distinct kinetic dependences of twofold and threefold fibrils show different properties of their respective frustrated complexes at the fibrils tips that were apparent as different kinetic parameters. Our findings illuminate the complex interplay between fibril structures and the associated transition states during the growth of amyloid β fibrils.

636890 The Interaction of β-Lactoglobulin with N,N-Dimethyldodecylamine N-Oxide and Sodium Dodecyl Sulfate

Abstract:

Surfactants are amphiphilic molecules that lower the surface tension between two phases. In aqueous solution, they are known to interact with proteins, forming protein-surfactant complexes which can affect protein structure, kinetic activity, aggregation behavior and binding to surfaces. As such, surfactants and mixtures of surfactants have been used with proteins for a variety of applications including formulation of cleaning agents, membrane protein extraction and separation, stabilization of biopharmaceuticals and foam formation. While the interactions of individual ionic and nonionic surfactants with proteins have been well characterized, in-depth study of surfactant mixtures in the presence of protein is still an emerging field. We expect combinations of surfactants to form mixed micelles that take on properties of each surfactant, resulting in altered interaction with proteins.

We used a variety of biophysical characterization techniques to investigate the behavior of mixtures of anionic surfactant sodium dodecyl sulfate (SDS) and amphoteric surfactant N,N-dimethyldodecylamine N-Oxide (DDAO) at varying molar ratios. Pyrene fluorescence showed that the critical micelle concentration (CMC) of surfactant mixtures is lower than the CMC of both pure SDS and pure DDAO, with little variation over a range of 0.10-0.65 mole SDS per mole total surfactant. Additionally, we investigated the interactions between DDAO-SDS mixtures and the model protein β-lactoglobulin (βLG). The surfactant concentrations at which micelles form were unchanged in the presence of βLG except above 0.85 mole SDS per mole total surfactant, where surfactant aggregates formed on the protein, similar to pure SDS. βLG is known to unfold in 2mM DDAO and 1.09mM SDS, yet circular dichroism and small angle X-ray scattering (SAXS) showed βLG was structurally stable in mixtures at higher surfactant concentrations. Simple linear combination of SAXS scattering curves revealed that βLG was most stable in mixtures of 75% DDAO and 25% SDS up to 5mM total surfactant. These results indicate that DDAO and SDS indeed form mixed micelles in solution with less favorable interactions with βLG which improves the stability of the protein in comparison to pure DDAO or SDS.

636469 Towards Improving the Accuracy of Protein-RNA Docking Predictions Using Hot Spot Residues: A Case Study

Abstract:

Protein-RNA interactions occur during, and are essential to, all stages of the Central Dogma, from DNA replication to expression of the phenotype. In addition, these interactions are becoming increasingly important through the design of novel biomedical devices and therapies which rely on protein-RNA binding. For example, investigators have developed RNA aptamers that bind protein biomarkers on Ebola, Ricin, and COVID-19, and they are actively investigating using aptamers as "chemical antibodies" in low-cost diagnostic devices and next-generation therapeutics. To study these interactions, researchers first need to obtain a three-dimensional structure for the protein-RNA complex. However, methods for doing so are currently limited. While experimental approaches are powerful, they are too laborious to keep up with the profusion of protein-RNA complexes that exist in nature or are generated by the development of RNA aptamers. While in-silico complex prediction methods could potentially alleviate such a bottleneck, as it currently stands these methods are unreliable in the absence of experimental data to guide and inform predictions. In fact, rigid-docking, a conventional in-silico approach for protein-RNA structure determination, is inherently limited by its treatment of protein and RNA molecules as rigid. It is well-documented in the literature that RNA molecules are flexible and undergo considerable conformational change upon binding. Here, we aim to address the current structure determination bottleneck by developing a workflow for improving the accuracy of in-silico structure predictions. The proposed workflow combines high temperature molecular dynamics simulations and manual formation of high-energy π-interactions with hot spot amino acid residues. A key benefit to this approach is it allows for the incorporation of RNA conformational changes, thereby addressing a current gap in the literature. Using a case-study approach, we evaluate the efficacy of the proposed workflow using the Critical Assessment of PRediction of Interactions (CAPRI) competition metrics and show that while the optimization workflow worsened the fraction of conserved native contacts(fnat) and ligand root mean square deviation (lRMSD) compared to the unoptimized starting prediction (fnat: 0.30 vs. 0.43; lRMSD: 10 vs. 8.1 Å), the iRMSD improved slightly (3.8 vs. 3.9 Å). Thus, while the efficacy of the optimization workflow is still undetermined, we believe that the underlying notion behind the workflow shows promise and could potentially be used to guide docking methods in the near future. In this presentation, we will discuss the current limitations of our proposed method and ways in which further research could help alleviate these issues.

636655 Towards an Alternate Psilocybin Biosynthesis Pathway Starting with Methylation of 4-Hydroxytryptophan

Abstract:

Psilocybin, the psychedelic compound found in a variety of bioactive mushroom species, has been used in clinical trials to study its efficacy towards the treatment of mental health disorders such as depression, anxiety, and post-traumatic stress disorder. The *in vivo* production of psilocybin follows a 4-hydroxyindole supplemented, three-step pathway that includes the enzymes psiD, psiK, and psiM. PsiM is a methyltransferase responsible for the conversion of norbaeocystin to baeocystin, and baeocystin to psilocybin by means of the methyl donor S-adenosyl-L-methionine (SAM). However, another SAM dependent N-methyltransferase derived from the *Psilocybe serbica* mushroom, TrpM, could lead to an alternative psilocybin production pathway that excludes the norbaeocystin methyltransferase, psiM. This novel pathway begins with the iterative methylation of L-tryptophan and results in mono-, di- and trimethylated products that could provide a possible substrate to psiD and psiK. In order to test this, the *in vivo* activity of TrpM needed to be examined. An *E. coli* vector containing the TrpM gene was put through a well plate assay resulting in the expected mono-, di-, and trimethylated products. Using 4-hydroxyindole as a supplement, TrpM was also able to carry out its processes on 4-hydroxytryptophan. This resulted in mono- and dimethylated products. This proposed alternative pathway begins with TrpM due to previously published findings that TrpM and psiM are not interchangeable; this is due to the different substrates that they accept. Once TrpM methylates 4-hydroxytryptophan, psiD can decarboxylate *N,N*-dimethyl-4-hydroxytryptophan, forming psilocin, and psiK can then phosphorylate to obtain the final product of psilocybin. Our efforts to evaluate this alternate biosynthesis pathway towards psilocybin will be presented along with analytical LCMS analysis demonstrating the first report of enzymatic *N*-methylation of 4-hydroxytryptophan *in vivo*.

637884 Controlled Biofouling of *Escherichia coli* on Perfluoroalkoxy Alkane in a Water Purification System for Testing Separation Performance

Abstract:

An emerging membraneless water purification technology is diffusiophoresis-based separations. Diffusiophoresis is a phenomenon that induces directed separation of colloidal particulate via chemical ion gradients. Diffusiophoresis can be used to separate bacteria from water, but biofouling may occur within the system. Thus, assessing the impact of biofouling on system performance is a critical step for evaluating diffusiophoresis-based water purification. Here, we establish the parameters for intentionally fouling perfluoroalkoxy alkane (PFA) with *Escherichia coli* to consistently yield biofilms with a height of ~10 μm. Shear stress and growth times were varied to modulate biofilm growth. Biofilm height and structure were quantified using image analysis of confocal laser scanning microscopy images. We found that growing *E. coli* in tryptic soy broth for 12-hours at a shear stress of 0.02 Pa resulted in the formation of 10 μm tall *E. coli* biofilms on PFA. Additionally, we report how the fluidic environment of a diffusiophoresis water purification system affects the structure of pre-formed *E. coli* biofilms. This work establishes the parameters for the controlled biofouling of a diffusiophoresis-based separator to evaluate the extent that biofouling influences the water purification system performance.

638290 Predicting Cell Cycling in Redox Flow Batteries Using Low-Dimensional Models

Abstract:

With increasing adoption of renewable energy sources, redox flow batteries (RFBs) offer an attractive solution to resolving their intermittent and stochastic nature. Compared to traditional batteries (e.g., Li-ion), RFBs have distinctive advantages for grid storage applications, including decoupled energy and power, design flexibility, and scalability. However, cost and reliability continue to present major hurdles to widespread commercialization. To improve RFB viability, materials discovery and cell engineering efforts typically aim to enhance device performance while reducing capital costs. Nevertheless, complex tradeoffs between various material properties, operating conditions, and performance metrics frustrate the identification of optimal system designs. Better understanding these functional relationships may improve materials selection criteria and provide critical insights into future engineering efforts.

In this work, we developed a low-dimensional model for translating fundamental material properties to cell performance. Our model, which leverages generalized assumptions about the redox flow cell geometry, can rapidly measure long-term cell performance metrics, including the capacity retention and energy efficiency. Using this approach, we explore the role of several key processes(e.g., mass transfer losses, ohmic losses, species crossover, and homogenous decay) under galvanostatic operation for one-electron redox couples. Because of its robust framework and rapid computation time, this model can be applied to optimization schemes, including parameter estimation and techno-economic analyses, thus taking a step further towards practical realization of RFBs.

638321 Parameter Optimization for the Performance of a Reversible Fuel Cell

Abstract:

A reversible fuel cell is an electrochemical device that can be operated as a fuel cell (discharging mode) combining hydrogen fuel and oxygen gas to produce electricity in a clean manner, with the only byproduct being water. It can also be operated as an electrolyzer (charging mode) splitting water to produce hydrogen and oxygen gases that can be stored and used for various applications. State of the art fuel cells and electrolyzers typically work in an acidic environment, which imposes the need of expensive precious metals catalysts to expedite the reactions taking place in the oxygen and hydrogen side electrodes. In our lab we are working on a device similar to a reversible fuel cell that produces electricity and hydrogen peroxide (instead of water) in the discharging mode and produces hydrogen and oxygen (from hydrogen peroxide) in the charging mode. This device works in alkaline media, opening the possibility of using non-precious metal catalysts for the oxygen and hydrogen electrochemical reactions, which can significantly decrease the cost of the device. Moreover, electrochemically produced hydrogen peroxide is a very sought-after chemical used as a bleaching agent in many industrial applications, adding value to our device. As part of my research, I prepare electrodes for the hydrogen and oxygen reactions using in house-made catalysts materials, assemble cells using in house-made membranes, and test its performance for charging and discharging under simulated laboratory conditions. I focus on optimizing parameters to enhance the performance of the device during testing while guaranteeing results reproducibility.

638419 An Equilibrium Potential Analysis of Bi and Cu Modified Rechargeable MnO_2 Cathodes

Abstract:

There is an urgent societal need for safe, scalable, low-cost energy storage technologies capable of sustaining the power grid at times when renewable power sources cannot. Currently, the energy storage market is dominated by lithium-ion battery technologies which make use of flammable electrolytes that combust in response to thermal runaway. Rechargeable Zn-MnO_2 alkaline batteries have been suggested as a suitable fit for grid-scale storage. Zn-MnO_2 cells use safe, non-toxic, earth-abundant materials making for an inexpensive (<$100/kWh) and sustainable battery technology. Accessing the full capacity of an MnO_2 rechargeable cathode requires that the conversion of Mn^{4+} to Mn^{2+} be completely reversible. Intermediate formation of some forms of Mn^{3+} during battery operation can lead to an irreversible phase change of MnO_2 to the Mn_3O_4 spinel structure. Once the spinel forms, that active material can no longer engage in the desired electron transfer reaction that stores energy. Recent studies suggest that incorporating Bi^{3+} and Cu^{2+} ions into layered MnO_2 cathode materials suppresses spinel formation and increases cyclability. In this work, the galvanostatic intermittent titration technique (GITT) is used to monitor the equilibrium potential of Bi- and Cu-modified MnO_2 cathodes at different states of charge during cycling. By correlating these equilibrium potentials to different redox states formed within the electrode, we seek to better understand the role of Bi and Cu in enhanced performance.

638663 Lithium Ion Battery and Chme Car

Abstract:

Lithium-ion battery optimization of active material is crucial due to the materials utilized in current lithium-ion battery technology. The cathode active material is typically constructed of nickel, manganese, and cobalt in varying molar percentages, but rich Nickel cathodes, such as NCM 523 and NCM 811, has been attracting increasing attention. However, high ratios of Nickel in cathode will occupy the cation mixing to decrease the battery stability performance. A novel and facile polymer assisted chemical solution is used to synthesize NCM material due to its ability to optimize structure and control particle morphology , especially at a nano-scale. Therefore, this method has the potential to increase capacitance, stability, and performance of a lithium-ion battery. Through the construction and testing of half cell batteries on either the Electrochemical Impedance Spectroscopy (EIS) or Battery Testing System we will be able to to examine the performance and determine how to improve the process.

The New Mexico State University Department of Chemical Engineering developed a chemical engineering competition car for the AICHE regional competition. The NMSU AICHE competition car was designed based on prior experience from the competition and the guidelines proposed by AICHE, which outlined that the car must chemically be powered and stopped independently of exterior assistance. NMSU had previously utilized a lead acid battery reaction to power the car and a trans cinnamaldehyde clock-based reaction to stop the car. The incorporation of research into lithium-ion battery would allow for a more efficient and successful storage of power to be utilized by the CHME car. However, implementing a lithium ion battery power system into the CHME car project poses several issues such as meeting required voltage, fabrication of batteries, and charge/discharge efficiency. Prior groups within the department have designed the car based on lead acid batteries for a multitude of reasons, namely that the design and implementation of lead acid batteries is easier, safer, more well understood than that of lithium-ion battery technology. Lithium-ion batteries must be assembled in a glovebox due to rapid oxidation of lithium and the electrolyte. The stopping mechanism was accomplished by a clock-based reaction of trans cinnamaldehyde and ethanol in varying proportions.

636888 Improved Efficiency of Lithium-Ion Batteries Using Electron Tunnelling

Abstract:

This article is majorly focused on two main sections:- (1) The first part of this article describes the formation of a Solid Electrolyte Interface which was considered as a blessing in disguise because it protects the electrolyte and suppresses further electrolyte reduction, but later it was discovered that this layer consumes 5% of total lithium ions present in the battery during its very first charge, but the heating of the seller can lead of melting of this layer and can cause an explosion in the battery. Electron Tunnelling a unique approach to decreasing this formation of SEI layer is proposed. (2) The second objective of this article is to simulate fully electric vehicles in MATLAB software. The simulation results are based on observations taken after considering necessary changes in the entire battery module, electron tunnelling and materials used in lithium-ion batteries.

638695 Scalable Aqueous-Based Synthesis of Graphene and Platinum Nanoparticle Aerogels for Electrocatalysis in Fuel Cells

Abstract:

Scalable Aqueous-based Synthesis of Graphene and Platinum Nanoparticle Aerogels for Electrocatalysis in Fuel Cells

Duncan R. Day, Uta A. Givens, Ryan N. Morrall, F. John Burpo, Caspar C. Yi, Enoch A. Nagelli*

Department of Chemistry and Life Science, United States Military Academy

West Point, New York 10996

ABSTRACT

Graphene/noble metal nanoparticle aerogels are promising conductive and ultra-high surface area three-dimensional (3D) electrodes that utilize the high capacitance of graphene in combination with the catalytic properties of noble metals for high performance energy storage devices. Their porous structure and conformability with other structural materials make them viable electrode candidates to be integrated into fuel cells and batteries. We report the successful development of a graphene-based aerogel formed first by the electrostatic assembly of platinum magnus salts with graphene oxide sheets. The subsequent chemical reduction of this assembled material results in a highly porous nanostructure that can enable electron transport between the graphene sheets and catalytic platinum. The aerogel is produced in a rapid and entirely aqueous process without addition of heat or cross-linking agents between the graphene sheets and platinum salts. After chemical reduction, the aqueous graphene-platinum solution is then critically dried to result in a monolithic solid structure that demonstrates high surface area within its porous structure. Scanning Electron Microscopy (SEM) is used to verify the integration of graphene sheets with platinum nanoparticles and nanotubes into an inter-connected 3D aerogel. The structural features of the aerogel was confirmed with Raman Spectroscopy with the presence of signature of D-band, G-band, and 2D-band peaks. Electrochemical impedance spectroscopy (EIS) was used to determine the specific capacitance of 442 mF/cm^2 and an areal specific resistance (ASR) of $0.41\Omega\ cm^2$. The high capacity and low electronic and ionic resistance of the graphene-platinum aerogel indicate the potential as catalytic electrodes in fuel cells.

KEYWORDS: Graphene, Platinum Nanoparticles, Fuel Cell, Electrocatalysis

CONTACT: **Dr. Enoch Nagelli**, **Dept of Chemistry and Life Science**, United States Military Academy, West Point, New York 10996. **Email: enoch.nagelli@westpoint.edu TEL:** 845-938-3904.

638047 Phase-Transitional Catalyst-Assisted Membraneless ZnI_2 Battery

Abstract:

Aqueous zinc-based batteries have been recognised as a promising solution for large-scale energy storage because of their safety, low cost, and high theoretical capacity. However, further development has largely been limited by a need for a membrane to eliminate self-discharge, which obstructs ion transport. This project has demonstrated a novel membraneless aqueous zinc iodide battery, using N-methyl-N-butyl pyrrolidinium iodide (MBPI) to complex molecular iodine released during charging into a quasi-solid charging product, which phase-separates from the bulk electrolyte, thus eliminating self-discharge. Moreover, this mechanism achieves increased iodide utilisation because iodine is not complexed by aqueous iodide ions. Raman studies have confirmed that polyiodide concentration in the electrolyte is significantly reduced in the presence of MBPI, and the MBPI charging product contains iodine. Electrochemical studies of cells containing 4 M ZnI_2 electrolytes have demonstrated 65 cycles with a Coulombic efficiency of over 85% with electrolyte containing 0.3 M MBPI, compared to 0 cycles without MBPI; the battery was cycled at a 100-mAh capacity, achieving an energy density of 18.6 WhL^{-1}, which exceeds many existing zinc-halide membraneless designs. Also, the presence of MBPI in the electrolyte inhibited Zn dendritic growth due to a reduced concentration of shuttling polyiodides in the aqueous phase, which cause electrolyte polarisation and produce a non-uniform electric field, thus initiating Zn dendritic growth. With its high capacity and specific current compared to existing alternatives, this phase-transitional membraneless zinc-iodide battery is a promising design for future energy storage systems.

637953 *Estimating Electron Transfer Kinetics for Flow Battery Electrodes Using Dense Carbon Films*

Abstract:

Redox flow batteries (RFBs) are promising technologies for the efficient and reliable delivery of electricity, offering opportunities to integrate intermittent renewable resources and to support unreliable and/or aging grid infrastructure. Within the RFB, the porous carbonaceous electrode provides surface area for redox reactions, distributes electrolyte, and conducts electrons. Understanding reaction kinetics of the electrode is crucial towards improving RFB output and lowering costs. However, reaction kinetics are driven by an interplay of factors and the complex geometries of porous electrodes invalidate the assumptions in conventional voltametric techniques used to assess electron transfer kinetics, thus frustrating our understanding of performance descriptors.

Here, we outline a strategy to estimate electron transfer kinetics on electrode materials reminiscent of those used in RFBs. First, we describe a bottom-up synthetic process to produce non-porous and planar carbon films to enable evaluation of electron transfer kinetics using traditional electrochemical techniques. Next, we characterize physicochemical properties of the films using a suite of spectroscopic methods. Last, we assess the performance of the films in a custom-designed cell architecture, extracting intrinsic heterogeneous kinetic rate constants in iron-based aqueous electrolytes using standard electrochemical methods (i.e., cyclic voltammetry, electrochemical impedance spectroscopy). We anticipate that the methods and protocols described in this work are broadly applicable for quantitatively assessing electrocatalysts.

638067 Development of NMC/LFP Cathode Material Blends for Lithium-Ion Batteries

Abstract:

The need for renewable energy is rapidly increasing, and the optimization of lithium-ion batteries is critical for the industry. Their high power and energy capability provides power for small consumables, such as watches, up to electric vehicles (EVs) and grid scale energy storage systems. Lithium-nickel-manganese-cobalt-oxide (NMC) is a favored cathode active material in EVs because of its high energy density and high voltage limits, with the challenge of inferior high rate performance than other chemistries. Lithium-iron-phosphate (LFP) is less common due to its lower voltage range but exhibits higher safety and better high rate performance, especially during pulses of high current demand. For this reason, blends of NMC and LFP can impart the high energy density of NMC, with the safety and pulsed performance of LFP.

The research conducted focuses on testing cathode materials by performing electrochemical and rate testing followed by electrochemical impedance spectroscopy (EIS). This experiment focused specifically on the cathode by assembling half-cells, which cycle the cathode against pure lithium. This allows any performance effects from the anode to be eliminated. The coin cells tested had NMC, LFP, or a blend of NMC/LFP as the primary material in the cathode. Blended cathodes were made with 70% NMC and 30% LFP. The EIS analysis allowed cell kinetics and transfer mechanisms to be assigned and compared between the two chemistries. The analysis attributed electrochemical features from both species, including the high voltage profiles from the NMC and the good pulse performance from the LFP.

637846 Performance of III-V Photoelectrochemical Devices in Neutral Electrolyte

Abstract:

Photoelectrochemical (PEC) cells separate water molecules into oxygen and hydrogen gasses using sunlight to extract hydrogen for fuel or feedstocks. Tandem III-V photoelectrodes, used in PEC devices, are known to have high efficiency but low durability. They are normally placed in acidic conditions and last between a few minutes to 10 hours. An acidic electrolyte is hazardous on a large scale and corrodes electrodes quickly, creating a barrier for commercialization. This work evaluated safer, neutral buffers for their potential to extend device lifetime. Nafion membranes and separators were also evaluated to separate oxygen and hydrogen gas products to prevent explosive mixtures of hydrogen and oxygen. Both Nafion and porous separators isolate products but were found to require different operating schemes. When using Nafion, a pH gradient built up between the anolyte and catholyte which decreased the performance by creating a greater overpotential for water splitting. When using a porous separator, a pH gradient was not observed, but their ability to prevent dissolved gas crossover requires further investigation. Using neutral electrolytes and a membrane in a PEC cell allowed for approximately a 10x increase in lifetime but at a decreased efficiency, which may be recovered with additional device and component optimization.

This research was supported by the Department of Energy: Office of Science as part of Science Undergraduate Laboratory Internship program.

638576 Material Attributes Impact Flow-through Pretreatment of Air-Classified Corn Stover: Experiment and Modeling

Abstract:

Biomass deacetylation is an important step in price-competitive industrial ethanol production. While this process usually employs batch reactors, previous studies have proposed using a fixed-bed continuous-flow reactor to increase process efficiency. Similar processes have been proved viable for woody biomass at several scales in the pulping industry. However, applying this process to herbaceous biomass is challenging due to its significantly higher compressibility which can cause flow-induced pressure drop across the solid biomass material in the reactor. To decrease the associated risks of scale-up for flow-through herbaceous biomass deacetylation, further studies are needed to analyze the effect of feedstock attributes on critical parameters including column pressure drop and permeability. In this project, eight corn stover types, air-classified uniquely by moisture level, particle size, and feedstock purity, were prewashed in a fixed-bed, lab-scale reactor and compacted at pressures ranging from 0 to 60 PSI. Pressure drop was measured for varying flow rates up to 5 liters per minute (LPM). Permeability and particle specific surface area were predicted using non-linear empirical methods based on Darcy's Law and the Kozeny-Carman Equation for fluid flow. Fixed-bed permeability was shown to be inversely and logarithmically proportional to compaction pressure for all sample types. Samples with higher moisture levels and larger particle size had a greater overall permeability, thus demonstrating stronger resistance to instrument-induced pressure drop. Feedstock purity did not conclusively influence permeability. In future fixed-bed deacetylation studies, these preferred feedstock types should be used to minimize pressure drop in order to reduce the chance of reactor shutdown due to bed blockage.

638682 Solar, Wind, and Thermal Storage in North-Central West Virginia

Abstract:

Solar, Wind, and Thermal Storage Integration

in North-Central West Virginia

Kevin B. Donnelly[1], Nagasree Garapati[1], Fernando Lima[1], Yingqi Zhang[2], Pierre Jeanne[2]

[1]*Department of Chemical and Biomedical Engineering, Benjamin M. Statler College of Engineering and Mineral Resources, West Virginia University, Morgantown, WV 26506.*

[2] *Lawrence Berkeley National Laboratory, Berkeley, 1 Cyclotron Rd, Berkeley, CA 94720*

The north-central region of West Virginia hosts unique geographic and geological opportunities for the development of integrated renewable energy systems with thermal storage. Currently, the state does not contain large-scale electricity facilities from solar power but does host several wind farms throughout the north-central region. Additionally, this region also demonstrates higher rates of heat flux than the surrounding Mid-Atlantic region. This work features the results of simulations done for wind and solar energy generation in the state and potential storage through a subsurface geothermal reservoir. Wind and solar power within West Virginia were simulated using System Advisor Model (SAM) from the National Renewable Energy Laboratory (NREL). Wind power production is validated by comparisons to current facilities in the state, while solar production is compared to a solar facility in Garrett County, Maryland. Power production data is utilized over the course of months for easier translation to the reservoir simulation. The subsurface reservoir was modeled using TOUGH2/EOS1, a program specifically developed to simulate coupled transport of water and heat in geothermal reservoirs. Reservoir behavior (pressure, temperature) over incremental years and decades of production time is analyzed. In order to achieve a more accurate pressure profile at well bores, a radial mesh is used around injection and production well bores in the simulations. The reservoir behavior is analyzed through temperature and pressure contour maps created at various time periods within the allotted timeframes. The preliminary results demonstrate the potential of such a system within the state, and the plan for future economic research is explored in this work.

Acknowledgment: This material is based upon work supported by the U.S. DOE's Office of EERE under the GTO, under Award Number DE-EE0008105.

638596 Rdf, an Alternate Energy Source for Cement Industries: Produced from Hazardous Wastes and Biomass Mixture

Abstract:

Cement industry is one of the prominent industries in Bangladesh. It is one of the fastest-growing industries in Bangladesh. In recent years, per capita cement consumption has become 187 kg, and there are almost forty government and non-government companies who produce cement. In our country, clinker, the main raw material of cement is not available. Thus, almost 30% of the production cost of cement is utilized to import clinker. The second major cost comes from power consumption, which is about 28% of the total cost. Power supplies in cement industries take place by burning coal. According to a report, in cement industries, about 30-35 kW electricity is required per metric ton, which is rising gradually. Due to burning lots of coal, a huge amount of CO_2 emission takes place. CO_2 emission is hazardous for the environment. They enhance respiratory problems caused by smog and air pollution, as well as contributing to climate change by trapping heat. Again, there are lots of hazardous wastes and biomass which get eliminated from rubber industries, textile industries, rice mills, wood industries, etc. such as - rice husk, saw dust, rubber granules, waste water sludge, empty fruit bunch, etc. These waste materials are not dumped or recycled properly sometimes, which causes pollution to water and soil.

Thus, this project is focused on developing an alternate energy source utilizing the hazardous wastes and biomass, which will be applied in cement industries. This alternate energy source is completely produced from waste materials, named as Refuse Derived Fuel (RDF). The production cost of RDF is very low, almost half of the price of the coal. Necessary calculation has been performed to determine the calorific value of RDF produced from available waste materials in our country, which is 22451 kJ/kg, which is almost similar to the calorific value of coal which is 23,000 kJ/kg. The utilization of RDF can decrease coal consumption up to 5%. Also, an economic analysis says these waste materials can decrease the cost of power consumption up to 40%. Allover, RDF is a very promising and amazing alternative for coal in cement industries considering its environmental, economic and financial impact.

638683 Plastics to Fuel: Potentials of Pyrolysis

Abstract:

The objective of the Plastics to Fuel project is to use pyrolysis to mitigate the burden of plastics pollution in rural and island areas. In 2018, only 8-10% of post-consumer plastic waste was recycled. By utilizing pyrolysis, a variety of commodity plastics can be converted into a usable diesel like fuel. Before plastics can be put into the reactor, the plastics are characterized for plastic types and any potential fillers using differential scanning calorimetry (DSC) and thermogravimetric analysis (TGA). Modifications of catalysts and the reactor bed have recently resulted in product yields of 80-90%. The quality of product is verified using Gas chromatography and compared to commercial and farming diesel standards. Plastics that have been run include PE, PP& PS pellets alone and in mixtures to simulate a mixed plastic waste stream; recycled agricultural PE pots; ocean recovered PE fishing nets; and red bag waste intact needles and syringes. All produced useable diesel product. The future involves redesign of the pyrolysis reactor for improved efficiency and recycle of the non-condensable gases for reactor heating.

637261 Effect of K+ Doping on Pyrolysis of Sawdust Based Biochar

Abstract:

Biochar can be used as a form of carbon sequestration, as a soil amendment, or as an alternative fuel source. Biochar is made by thermally decomposing a plant waste-based feedstock in an inert atmosphere at high temperatures in a process called pyrolysis. Prior experiments established a correlation between inorganic content and relative mass yield of biochar and had results which contradicted each other. In preexisting studies, ash content was adjusted by selecting different plant-based feedstocks with known ash content values. In this study feedstocks were doped with inorganics (1-5% K^+) to minimize the impact of non-independent variables on biochar yield. The experiment reveals a strong positive correlation between K^+ content and final relative mass at the end of the pyrolysis reaction as well as a lowering of the reaction temperature by 70°C. The increase in biomass yield may be caused by an alternative reaction pathway enabled by the catalytic effect of K^+ ion or the OH^- ion reacting to increase the porosity of the sawdust and increasing the surface area of the overall reaction.

637630 Role of Hydrogen Gas in the Degradation of Natural Gas Pipelines

Abstract:

In response to rising greenhouse gas emissions, the demand for more sustainable fuels has greatly increased. Hydrogen (H2), a completely decarbonized gas molecule, has been identified as one potential replacement for natural gas (CH4). Using hydrogen gas can reduce carbon emissions but presents many challenges regarding safe transportation and storage. The purpose of this project is to determine the effect of hydrogen on the lifespan of existing natural gas pipeline infrastructure. To do so, multiple pipeline samples have been prepared and these will be exposed to pure hydrogen and mixtures of methane and hydrogen in a static low-pressure environment. The levels of corrosion and other chemical and physical changes in metal crystallinity in the metal samples is being quantified by mapping and using XPS, XRD, SEM, TGA, and DMA techniques. We are also studying the degradation of polyethylene pipeline samples. The methodology involves characterizing pipeline samples before and after exposure to hydrogen and/or mixtures with methane. The ongoing and planned measurements will determine the viability of hydrogen as an alternative to natural gas in metal and polymer pipelines.

637640 Simulation of an Interesterification Biodiesel Production Plant

Abstract:

Growing concern regarding automobile emissions and transportation energy demands has increased the viability of biodiesel production for commercial use. Biodiesel commonly consists of fatty acid methyl esters (FAME) and is typically produced from triglycerides (TG) via a transesterification reaction with methanol. However, this reaction produces glycerol as a byproduct in quantities that are quickly surpassing the market need. Alternatively, the interesterification of TG with methyl acetate (MA) produces FAME and triacetin (TA). Unlike glycerol, TA is an effective fuel additive and does not need to be separated from the FAME mixture. TA can also be sold separately from the biodiesel as a plasticizer and gelatinizing agent for polymers and explosives. Interesterification occurs in three reversible reactions that require an excess of MA, or a high methyl acetate to oil molar ratio (MAOMR), to facilitate FAME production. Typically, heterogeneous or homogenous, acid or base catalysts are used. Investigating interesterification is important considering its potential technical and economic advantages compared to transesterification. This work simulates a preliminary process model for an interesterification biodiesel production plant with a target production rate of 50,000 metric tons per year. The model was created in three stages: (1) modeling intermediate compounds in Aspen Plus; (2) obtaining and verifying published kinetic parameters using a heterogenous base catalyst; and (3) investigating separation techniques for biodiesel purification. This simulation will inform an ongoing techno-economic analysis of the biodiesel production plant.

637798 The Effect of Polymeric Additives on Biodiesel Blend Cold Flow Properties

Abstract:

Biofuels prove to be a useful alternative to petrol diesels, as they can be created from naturally sourced materials like animal fats, plant oils, and cooking oils. However, they currently face issues concerning solidification in cold environments. The cloud point (CP) signifies when different components of the biodiesel begin to crystallize, causing solidification. By reducing the CP, the overall cold flow performance of the biodiesel can be improved. Studies have shown that the CP of biodiesel can be altered by doping the fuel with petrol diesel or wax-inhibiting polymers dissolved in solution. This research aims to study the effect of mixing ratios of diesel, biodiesel, and polymers on the CP. Various blends of biodiesel and petrol diesel with either 1 wt% or 5 wt% polymer additive were created. Solutions were analyzed via differential scanning calorimetry. Results showed that as the amount of biodiesel in solution increased, the CP decreased. For similar blends, as the weight percentage for both additives increased, the CP slightly decreased. The PEG solutions performed better than the EVA solutions before the B50 blend ratio, but the EVA solutions performed better after the B50 ratio. This result is likely due to the relative sizes of each polymer and biodiesel and diesel structure. EVA has a more complex structure than PEG, so it will have a harder time crystallizing with the complex biodiesel structures.

638154 Mixed Photosystem I and Cytochrome c Films for Biohybrid Solar Energy Conversion

Abstract:

The Photosystem I (PSI) protein can harness solar energy to facilitate redox reactions and has been utilized extensively in the field of biohybrid solar energy conversion. Improvements to the performance of such PSI-based biohybrids are dependent on efforts to engineer the electron transfer pathway between the reaction centers of PSI and the electrode. In this work, we use the fast and versatile vacuum-assisted drop-casting technique to develop a multilayer PSI photocathode incorporating cytochrome c (cyt c) as an electron carrier. Cyt c is entrapped near the surface by the PSI assembly and perfluorosulfonic acid/polytetrafluoroethylene (Nafion) ionomer top layer with areal densities of 21 ± 16 pmol/cm^2 as measured via cyclic voltammetry. The photocurrent generated by this system with oxygen as an electron acceptor at a negative applied potential is 240 ± 115 nA/cm^2, a marked improvement compared to 31 ± 19 nA/cm^2 generated by a PSI multilayer assembly without cyt c.

639078 Hydrothermal Liquefaction of Hazardous Biomass

Abstract:

Hydrothermal liquefaction (HTL) is one of the many ways to convert biological feedstocks into liquid fuels that can be applied in various ways. The HTL process utilizes water chemistry at high temperatures and pressures to produce the feedstock into three distinct phases. Using food waste as a feedstock has the potential to reduce the emissions from landfills. Food waste that is processed using HTL can be implemented in compost production due to the high concentrations of nitrogen and water. Another feedstock processed was radioactive algae, with the goal of tracking the phase in which the uranium was transferred to post-conversion. Sewage sludge was also investigated as a feedstock. The HTL process eliminates blood-borne pathogen hazards, making it valuable for real world applications. The nutrients recovered can be used for fertilizer. The current project studies the HTL of resin and bagasse from the guayule shrub. After processing, there are three products: rubber, resin, and bagasse. Using the HTL process, imitating cracking hydrocarbons and removing oxygen containing compounds can be achieved, liquid fuels can be produced and separated. The three phases will be characterized and analyzed for the properties of interest. A known market for these fuels is the jet fuel industry.

638055 Styrene Sulfonate As a Surfactant-Monomer (surfmer)

Abstract:

High internal phase emulsions (HIPEs) are highly concentrated systems resulting from the mixing of two immiscible liquids with a surfactant (emulsifier). A polyHIPE is obtained from the polymerization of these HIPEs to create porous polymers. Both systems have been growing areas of research, having applications in tissue engineering, smart materials, water treatment, and reaction support, to name a few. Based on today's water quality concerns, new and affordable ways of water treatment are desired; precisely these kinds of treatments represent another polyHIPE application. The goal of the research focuses on the inclusion of another family of surfactants; here, styrene sulfonate (SS) is introduced as a candidate, acting as a surfactant-monomer compound (surfmer). Surfmer compounds are desired for their combination of the physical behavior of surfactants and the reactivity of monomers which permits further chemistry added on the product. For it to qualify as a surfactant, it needed to lower the interfacial tension between immiscible phases. This step was performed and confirmed using pendant drop tensiometry to record the interfacial tension at the interface. Later on, porous beads will emerge from this polymerization process. Lastly, HPLC (high performance liquid chromatography) tests were used to confirm full incorporation of the studied surfactant-monomer in the product. The goal is to obtain a minimal concentration of surfmer in the effluent lowering this value as much as possible down to a targeted value of 0. In the end, optimizations in the formulation process are recommended to obtain better porous polymer production.

638573 Development of Clay Nanocomposite Hydrogels for Pfas Remediation

Abstract:

Environmental contaminants can be extremely harmful to humans and animals in high concentrations within the body. Per-and polyfluoroalkysubstances (PFAS) are a class of environmental contaminants that are considered forever chemicals because of the difficulty for it to degrade in the environment, which makes it a very dangerous substance when it accumulates in the human body and environment. Enterosorbents are a class of natural and synthetic materials that are orally administered and pass through the GI tract removing harmful contaminants. Clay is a sorbent material that has proven to bind a range of molecules and transport them without displaying toxic effects within the body. It can also be incorporated into a hydrogel, a hydrophilic crosslinked network, to increase the hydrophobic nature of the material for greater binding affinity. A series of acrylamide-based hydrogel nanocomposites were developed with montmorillonite clay to remove PFAS from the human body. The crosslinker N,N'-methylene-bis-acrylamide (NNMBA) was used, and the initiator ammonium persulfate (APS) and activator tetramethylethylenediamine (TEMED) allowed for free-radical polymerization to occur. The use of crosslinked hydrogels as enterosorbents and the ease of functionalization allow for highly selective materials with affinity towards specific compounds. Values of 0.1, 1, 5, and 10 mole percent of the crosslinker to the total monomer hydrogels and nanocomposites were developed. Swelling studies were performed on all systems with the neat (without clay) polymers having a swelling ratio of 8.07 ± 2.22 (0.1% NNMBA), 5.69 ± 0.06 (1% NNMBA), 3.06 ± 0.05 (5% NNMBA) and 2.55 ± 0.05 (10% NNMBA). As the percentage of crosslinker (NNMBA) increases, the swelling ratio decreases, and this is consistent across both the 1% and 5% weight loading of clay. Future sorption studies utilizing perfluorooctanoic acid (PFOA) and perfluorooctane sulfonate (PFOS) will be performed to determine the binding capabilities of these enterosorbents.

638435 Ion-Exchange Chromatography for the Separation of Critical Elements from Bioleachate

Abstract:

Lithium, Manganese, Cobalt, and Nickel are essential materials in the production of batteries and circuits. Global demand for these metals will only grow as electric vehicles and electronic devices become ubiquitous. The bulk of these materials are imported from foreign entities; supply is insecure and high-risk.

In an effort to develop a sustainable, domestic source of these critical materials, our group is cultivating a process to recover Lithium, Cobalt, Nickel, and Manganese from recycled lithium-ion batteries (LIB). This process involves leaching the recycled LIB in a bacteria-produced acidic solution, yielding a bioleachate rich in metals. These metals must then be isolated from the bioleachate. This project aimed to accomplish this separation of metals by way of ion-exchange (IX) resin chromatography. Specifically, Dowex-M4195 was tested for its purported ability to separate Cobalt and Nickel from Lithium and Manganese. Ultimately it was found that a mixture of six primary elements in bioleachate- Ni, Cu, Co, Fe, Mn, and Li- could be separated into three fractions using a column packed with Dowex-M4195. The first fraction, the raffinate, contained Mn, Li, and Fe(II). A weak sulfuric acid strip isolated Fe(III) and Co. Finally, a strong sulfuric acid strip removed Ni and Cu and regenerated the resin. The column, once rinsed with water, could be once again loaded with bioleachate.

638516 Modeling of Diffusion in the Metal-Organic Framework Uio-66

Abstract:

Highly porous metal-organic frameworks (MOFs) have been studied for the capture and destruction of chemical warfare agents (CWAs), with the UiO-6x family showing great promise. This work explores UiO-66 specifically using molecular dynamics. Transport phenomena are especially important in understanding the behavior of CWAs in UiO-66. To explore diffusion in UiO-66, isopropanol (IPA) has been chosen as the primary analog molecule due to its ability to hydrogen bond to hydroxide groups within the MOF, like many CWAs are expected to, and its relatively rapid diffusion in pristine UiO-66, unlike most CWAs, which are only expected to diffuse appreciably in defective UiO-66. Self and corrected diffusivities were examined, with the latter used to calculate transport diffusivity by application of a thermodynamic correction factor. Diffusion free energy profiles of several additional molecules of relevance, including notable CWA simulants dimethylphosphite (DMP) and dimethyl methylphosphonate (DMMP), were calculated. Hydrogen bonding was found to play a substantial role in the diffusion of polar species, with an order of magnitude decrease in diffusivity and a twofold increase in the activation energy to diffusion of IPA when hydrogen bonding was properly accounted for in the parameterization of the system. An understanding of base interactions in the pristine MOF serves as a starting point for exploring more complex defective structures and a wider range of guest species in future studies.

638344 Self-Movement of Liquid Drops on Swollen Polydimethylsiloxane Films of Varying Thickness

Abstract:

Self-movement of a liquid drop on a surface is controlled by surface tension forces, which can be quantified by a difference in contact angles between two "sides" of a drop. Differences in contact angles are typically induced by chemical, electrical, or topographical gradients. However, swelling and thickness gradients on a soft surface may also induce drop motion. When a drop is placed on soft and swollen elastomers, network deformation and potential fluid separation occur. This means that there exist both pull-up of the polymer network (a solid wetting ridge) and pull-up of the swelling solvent (an oil ridge) at the periphery of the drop. The heights of these ridges are determined by a balance of surface tension of the drop and restoring forces of the polymer network, which thus impact contact angles. Here we study the dependence of contact angles on the elastomer film thicknesses, the degree of network crosslinking, and the degree of network swelling using polydimethylsiloxane (PDMS) swollen with silicone oil. Above a certain swelling ratio, we find that the contact angles for 4 μL water drops noticeably decrease as film thickness increases within a range of 1-250 microns. Our results suggest that a thickness gradient imparted to highly swollen elastomer films can promote spontaneous movement of water drops without the need to apply external energy.

638232 Viscosity Scaling of Diallyl-Dimethyl-Ammonium Chloride Polyelectrolytes in Aqueous and Salt Solutions

Abstract:

A better understanding of the different concentration regimes that exist for polyelectrolytes can be helpful to create membranes for anion exchange with applications including electrolysis and fuel cells. Viscosity scaling is a method used to study various concentration regimes in polymer and polyelectrolyte solutions. The objective of this study was to perform the viscosity scaling for two cationic polyelectrolytes in water and 0.1 M NaCl: a co-polymer Poly(acrylamide-co-diallyl-dimethyl-ammonium chloride)(PAAcDMAC) and a homopolymer Poly(diallyl-dimethyl-ammonium chloride) (PDADMAC). Concentrations from 0.25 to 18 wt% polymer spanned from dilute to entangled concentration regimes. Higher concentrations provided insight on the semi-dilute and entangled regimes, which have not been extensively studied. Comparing the PAAcDMAC in 0.1 M NaCl in the dilute region, 0.25 to 1.0 wt%, the ions from the salt interact and screen some of the electrostatic interactions. As a result, the polymer assumed a more compact conformation and viscosity decreased. Thus, in DI water, the electrostatic interactions of PAAcDMAC repel each other and cause the chain to elongate. This finding confirms previous studies on polyelectrolyte rheology. At concentrations above the dilute regime, the effects of salt solution versus water decreased until little difference in viscosity existed for PAAcDMAC. PAAcDMAC in DI water encompassed all four concentration regimes, and PAAcDMAC in NaCl exhibited three concentration regimes. Similar concentration regimes were measured for PDADMAC solutions. Beyond flow viscosity measurements, the viscoelasticity was measured using small amplitude oscillatory shear and large amplitude oscillatory shear.

638285 Sustainable Polylactic Acid Foam Development Via Solid-State Processing and Cross-Linking

Abstract:

Plastics have played a vital role in global materials industries for the past 80 years, but their contribution to climate change and their inability to biodegrade when disposed of the environment has generated great cause for concern. Bioplastics have replaced some of the traditional petroleum-derived materials over the recent years, with room for a wider range of applications and extensions. One important category of plastic material is polymer foams, which have widespread applications in insulation, packaging, and storage. The work covered in this project aims to generate, test, and analyze bio-based polylactic acid (PLA) polymer foams that undergo solid-state processing, such as solid-state shear pulverization (SSSP) and cryogenic milling, as a pre-process prior to foaming. Solid-state processing has been shown to enhance crystallization kinetics of PLA, which theoretically enables better PLA foaming by allowing PLA to crystallize quicker and therefore capture more gas. This study focuses on foaming through a chemical foaming method, using an exothermic, nitrogen-based chemical foaming agent (CFA). Melt extrusion and compression molding are used to generate the foams after initial solid-state processing. It is determined whether the enhanced properties through solid-state processing support chemical foaming to a sufficient extent to enable PLA foams that are comparable with those that use petroleum-derived cross-linkers such as dicumyl peroxide (DCP) and tryallylisocyanurate (TAIC), which are currently the industry standard.

638304 Metal Organic Framework - Fabric Based Filters for Particulate Matter Filtration Using Large Scale Production Methods

Abstract:

Metal-organic frameworks (MOFs) have been utilized as adsorbents and filters for a wide range of materials, including gases, small molecules, proteins, etc. MOFs can be utilized as particulate matter filters if the MOFs are integrated into fabrics other polymeric membranes. However, the scalability of creating MOF-fabrics to obtain large area, high throughput production is one of the significant issues preventing a widespread adoption of MOF-fabrics for particulate matter filtration. Chiefly, there are few simple methods of obtaining rapid MOF growth in a manner suitable for large area MOF-fabric production. Here, we show a novel method of rapidly synthesizing the prototypical MOF UiO-66 on different fabrics using multiple cycles of sequential dip coating and the utility of these MOF-fabrics for particulate air filtration. We show that UiO-66 can be integrated into the fibers of multiples fabrics, with varying levels of efficacy. From these, we take the best-performing and industrially relevant fabrics, cotton and polyester, and study the impact of performing dip-coating using an industrial roller coater to be utilized in the future in roll-to-roll coating processes. These MOF-fabrics show continued efficacy even after 25 washing cycles. These results show reusable, rewashable MOF-fabrics for air filtration can be created using industrial roll-to-roll coating techniques.

638148 Fabrication of Guest-Host Interacted MOF Thin Films Using Solution Shearing

Abstract:

Most metal–organic frameworks (MOFs) have an insulating nature due to their porosity and redox-inactive organic components. The electrical conductivity of the prototypical MOF, HKUST-1, can be tuned by infiltrating a small-molecule organic semiconductor, 7,7,8,8-tetracyanoquinodimethane (TCNQ), into the HKUST-1 pores, creating TCNQ@HKUST-1. However, current processes of creating TCNQ@HKUST-1 films have many roadblocks such as slow crystallization rates, which limit high throughput production, and the formation of Cu(TCNQ) as a byproduct, which affects the electrical conductivity and degrades the chemical structure of HKUST-1. In this work, we show that HKUST-1 films can be rapidly synthesized over large areas with consistent thickness and no pinholes via a meniscus-guided coating technique called solution shearing. The subsequent pore activation process and TCNQ impregnation can be completed via solvent exchange to minimize the formation of the Cu(TCNQ) byproduct, and we obtain an increase in electrical conductivity of the solution-sheared TCNQ@HKUST-1 thin films of over 7 orders of magnitude, reaching a maximum value of 2.42×10^{-2} S m^{-1} when TCNQ is incorporated for 10 days. The conductivity of solution-sheared TCNQ@HKUST-1 is higher compared to films formed by high-pressure pelletization of TCNQ@HKUST-1. We show that solution shearing can produce large-area thin films rapidly and reduce the formation of grain boundaries better than pelletization, allowing for large-area electronics with both charge transport and porosity for applications as sensors and electronics.

637417 Molecular Dynamics Evaluation of Differing Pore Geometry in Nanoporous Membranes for Reverse Osmosis Desalination

Abstract:

Nanoporous membranes based on nanoporous graphene or other emerging classes of materials have shown huge potential in reverse osmosis (RO) water desalination for their high throughput and excellent salt rejection. Recently, a number of promising RO membrane candidates have been proposed, but a closer understanding of the mechanisms behind variations in performance of these materials is needed for their further selection and development. The geometry of the pores in these nanoporous materials has proven critical to their ultimate performance, so it is desirable to understand the effect of various pore geometries and the nature of their interactions with saline water. This project aims to develop a widely applicable interpretation to this problem by testing a generalized toy model with varying representative pore shapes using molecular dynamics. Simplified pores exhibiting roughly circular, square, and triangular geometries with a range of pore sizes have been tested to identify differences in key performance markers of flux and salt rejection. Additionally, the use of molecular dynamics enables an atomistic understanding of the separation process through each pore shape, and has provided insights into the distribution and movement of water molecules and salt ions within and around the pores. This information about relative efficacy of differing pore geometries as well as insight into the dynamics of individual particles within the pores will be useful in streamlining and informing future study of nanoporous materials for improved RO membranes.

638415 Development of Cationic Hydrogels for Pfas Remediation

Abstract:

Per- or polyfluoroalkyl substances (PFAS) are a group of man-made chemicals found in packaged food, drinking water, and common household products. As a result of bioaccumulation from long elimination half-lives, PFAS are detrimental to human health, with consequences ranging from hormone disruption to cancer. Current methods of PFAS remediation rely on sorption in neutral environments, where these contaminants exist in an anionic form. Hydrogels are crosslinked, hydrophilic polymers that swell in aqueous environments. Binding of PFAS is impacted by electrostatic and hydrophobic interactions, as well as the sorbent morphology; therefore, the aim of this study is to develop cationic hydrogels to maximize anionic organic contaminant selectivity. Acrylamide based hydrogels were synthesized via free radical polymerization with different densities of the crosslinker, N,N'methylenebisacrylamide (NNMBA). Varying crosslinker concentrations allow size selectivity by controlling the structure of the mesh network. To further increase the affinity for anionic contaminants, a range of cationic monomers were introduced to the hydrogel system. Polymerization and conversion were confirmed using FTIR spectroscopy. Swelling studies in deionized water provide insight to the expansion capacity of the polymer in these neutral conditions. As expected, an increase in the crosslinking density of the hydrogel restricts the hydrophilic network, and therefore is accompanied by a decrease in swelling. Additionally, a greater percentage of cationic monomer present in the hydrogel increases swelling as well because of charge repulsion and the osmotic effect. Future studies aim to develop these polymeric systems for PFAS elimination in humans through the gastrointestinal tract.

638657 Effects of Osmolyte Concentrations on Complex Coacervate Systems

Abstract:

Vaccine delivery has never been more important than it is now. Current vaccine delivery methods require significant refrigeration to keep a vaccine stable, making it difficult to vaccinate in locations that have minimal resources for deep freezing. Ideally, a vaccine should be able to be stored at room temperature. Complex coacervation can potentially increase vaccine stability at greater temperatures by encapsulating vaccine material and mimicking the microenvironment of that material. Coacervation is a liquid-liquid phase separation that occurs from the electrostatic complexation of charged polymers and the release of counterions resulting in significant gains of entropy. This phenomenon has been used to encapsulate biological materials, create films, and much more. Research on complex coacervation generally studies the coacervate phase behavior as a function of salt concentration. The addition of salts can suppress the interaction between polyions to inhibit counterions from releasing, lowering the entropic gains and eventually disfavoring phase separation. In a biological context, this would release the material contained within a coacervate, with a potential application being vaccine delivery technology. However, certain biological materials are denatured by the presence of such a sufficiently large salt concentration and injecting humans with such concentrations of salt can be highly irritating. These factors discourage salt usage in applications like vaccine delivery. We hypothesize the use of biofriendly osmolytes can help or even replace salt as a means of disassembling a coacervate. The osmotic materials that are of particular interest include sugars and amino acids, materials which already have uses in stabilizing vaccines. These possibly charge-neutral osmolytes have the potential to modulate coacervate phase behavior by altering the osmotic pressure balance in the system. We will measure the critical concentrations of osmolytes to understand their effect on phase behavior. Measured critical concentrations of these osmolytes can prove useful for future coacervate research that requires controlled biological material release, such as vaccine research.

638676 Nonfouling Thin Film Hydrogels

Abstract:

Bacterial adhesion to surfaces is an issue that can lead to infection and disease in individuals that encounter contaminated surfaces. To prevent this, procedures for coating multiple surfaces with thin film, nonfouling polymers are being developed. Nonfouling polymers are defined as materials that resist nonspecific protein adhesion and bacteria adhesion. A few known chemistries that present nonfouling properties are zwitterionic functional groups and polyampholytes. Zwitterions are molecules that possess both a positively and negatively charged region in their structure but maintain an overall net charge of zero. One example being used in this work is sulfobetaine methacrylate (SBMA). Polyampholytes are formed with mixtures of positively and negatively charged monomers and they behave similar to zwitterions when they maintain an overall net charge of zero. The main polyampholyte coatings included in this work are formed from [2-(acryloyloxy)ethyl] trimethylammonium chloride (TMA) and 2- carboxyethyl acrylate (CAA) monomers or TMA and 3-sulfopropyl methacrylate (SA) monomers. The primary objective of this study is to develop procedures for coating thin film polymer hydrogels on three different substrates: aluminum 6061, stainless-steel, and gold. Thin films are being pursued over the more widely studied polymer brush coating techniques due to the ease of coating large surface areas through this technique. The thin film coatings are being attached to the underlying surface chemistry through surface bound species that participate in the hydrogel polymerization reaction. The results will demonstrate the successful coating procedures developed for all three metallic surfaces.

638465 Macroporous Hydrogels for Soil Water Retention

Abstract:

Amidst the challenges of climate change and drought, water scarcity is an issue that is persistent in regions of semi-arid climate. These conditions negatively impact agricultural production. The goal of this project was to successfully increase water retention using a macroporous hydrogel that absorbs water more rapidly via capillary action. The macroporous hydrogels were templated using high internal phase emulsions made with cyclohexane and olive oil. These emulsions were then polymerized using the catalyst N,N,N',N'-tetramethyl-ethylenediamine (TEMED) and then air dried. These macroporous hydrogels were then integrated into sandy soil. The water holding capacity and capillary rise were tested to measure the available water content and water uptake rates. Different amounts of the hydrogel were incorporated within the soil in order to measure the effect that the presence of these hydrogels have on soil water retention. The treated soil exhibited significantly higher water retention than the untreated soil. The higher water uptake rate is due to capillary action, and the treated soil shows a higher maximum uptake due to porosity. A scanning electron microscope (SEM) analysis showed that a highly interconnected structure was formed within the macroporous hydrogel. The hydrogel investigated has displayed a higher water uptake and increased absorbance due to the internal macroporous structure when compared to current super absorbent polymers.

638487 Utilizing 3D Hydrogel Systems to Evaluate the Proliferation and Differentiation of Oligodendrocyte Progenitor Cells in the Presence of Electrospun Fibers

Abstract:

Demyelinated neural environments, such as those affected by multiple sclerosis (MS) or traumatic brain injury, require an increased presence of, and remyelination by, oligodendrocyte cells (OLs) to repair the damage. These OLs stem from lineage-restricted oligodendrocyte progenitor cells (OPCs), which can both proliferate and differentiate. The Lampe Group commonly encapsulates OPCs in extracellular matrix-mimicking 3D hydrogel systems to observe cell response to stimuli, where lower stiffness gels are conducive to higher rates of viability and proliferation[1]. We have observed that the addition of axon–mimicking electrospun fibers in norbornene-functionalized hyaluronic acid (NorHA) gels leads to increased OPC process extension, a preliminary indication of differentiation toward OLs. This contrasts the circular morphology of cells found in similar gels that do not contain fibers. Since both the proliferation and differentiation of OPCs are integral to the remyelination process, we aim to create environments to study and maximize each rate. In the current hydrogel system, we hypothesize that the energy devoted to morphology changes in the presence of fibers reduces the proliferation rate. Over a 7-day period, NorHA gels containing 1% methacrylated hyaluronic acid (MeHA) electrospun fibers and gels containing no fibers were compared using EdU and DAPI stains. When cells synthesize DNA during proliferation, the fluorescent EdU stain is incorporated into the new strands. DAPI is a fluorescent stain that binds to all DNA, fluorescently labeling each cell to give a total cell count. Ratios of the two stains were utilized to measure average proliferation per cell. OPCs encapsulated in gels without fibers consistently proliferated at higher rates than OPCs in the fiber-containing gels (roughly 2-fold higher, day 7), which instead extended more and longer branches. This indicates that the presence of axonal-like fibers increases OPC differentiation into OLs but also decreases proliferation. To confirm trends seen in the former analysis method and to ensure no technique bias, the ratio of EdU to DAPI signal area was used as an additional measure of proliferation rates. The results of this analysis also allowed the conclusion that proliferation decreased while process extension increased with the addition of fibers.

References:

1. Unal, D. B., Caliari, S. R., & Lampe, K. J. (2020). 3D Hyaluronic Acid Hydrogels for Modeling Oligodendrocyte Progenitor Cell Behavior as a Function of Matrix Stiffness. doi:10.1101/2020.04.01.020412

638499 Novel High-Throughput Characterization Method for Soft Material Bulk Mechanical Properties

Abstract:

In the current pipeline for novel materials discovery to widespread adoption, new materials can be synthesized far faster than they can be characterized. This undesirable bottleneck is largely caused by the limited throughput of current materials characterization methods. There are few existing high-throughput mechanical testing methods, as simultaneously applying the same mechanical force on many samples is difficult. Moreover, the existing methods are either too costly or complicated to perform on a sufficiently widespread scale. We aim to help alleviate this bottleneck with a novel centrifugation-based high-throughput characterization method for bulk mechanical properties of soft materials. Centrifugation is used to exert a uniform force on stainless steel microparticles that have been embedded in hundreds of different material samples on a well plate. These particles then exert a stress on the samples and eventually break them, allowing us to measure the maximum stress each sample can withstand. The data we obtain from this method shows strong correlation with standard indentation testing. One major benefit of this method is that it is simple to perform and requires only common laboratory equipment, such as well plates and a standard tabletop centrifuge. This technique can easily reach a throughput of thousands of samples per run, with the only limitations being the sizes of the well plate and the centrifuge. Due to the relative simplicity, low cost, and extremely high throughput of this method, we believe it could help to expedite the pipeline from discovery to widespread adoption of novel materials. In addition to overcoming the bottleneck of materials characterization, this method could potentially be utilized to generate massive databases of material properties. These databases could then be fed into machine learning algorithms to generate predictive models that can aid in the discovery and creation of novel materials with desired properties.

638256 Structure-Property Relations of Epoxy/PCL Based Shape Memory Composite Hydrogels

Abstract:

Shape memory polymers (SMPs) are a special category of polymers that undergo a temporary shape and recovery to a permanent shape when exposed to external stimuli such as heat or water. This makes them valuable in the medical device industry, specifically in areas such as endovascular applications, drug delivery and tissue engineering. Shape memory properties, in particular, allow for minimally invasive surgeries.

In this project, epoxy/PCL-based triple shape memory composites (TSMC) called polyethylene oxide-epoxy/poly(ε-caprolactone) or PEO-EP/PCL have been prepared using electrospinning. In order to create this TSMC a 20 wt% poly(ε-caprolactone) (PCL) solution dissolved in a 4:1 chloroform to DMF ratio was mixed for 24-36 hours, following which solutions were electrospun into thin fibrous mats. Subsequently, a miscible blend of diglycidyl ether of bisphenol-A (DGEBA) with Jeffamine ED2003 (2:3) was imbibed into the PCL fiber mat and cured to a pore-free translucent composite. Interestingly, because the ED2003 diamine features a water-loving poly(ethylene oxide backbone) the resulting composites swell in water, enabling formation of a composite hydrogel. We investigated the structure-property properties of the composites, including differential scanning calorimetry (DSC) to characterize the complex melting behavior of PEO-EP/PCL Further, dynamic mechanical analysis (DMA) was conducted to characterize the tensile storage and loss moduli, while tensile testing of the dry and wet forms of the composite enabled determination of ultimate mechanical properties of the unique materials. Shape memory properties were found to enable distinct heat/hydration recovery of the permanent shape from arbitrarily fixed temporary shapes.

638260 Optimizing the Degree of Methacrylation for Polyelectrolyte Complex Microcapsules

Abstract:

Osteoarthritis is a chronic degenerative joint disease that affects millions of adult Americans. The effects of this disease go beyond pain causing lifestyle changes and impacting economic well-being. Current leading treatments are not permeant solutions for patients and still hold a multitude of potential risks. One challenge of these treatments is cell growth and integration. An emerging solution to aid in the growth, handleability, and integration of chondrocytes is using polyelectrolyte complex microcapsules to house the cells. To optimize microcapsules for successful future use, this research investigated the effects of the degree of methacrylation on the capsule membrane for a range of methacrylic acid methacrylated hyaluronic acid (HA-MA) ratios. This range included 1:0.1 HA-MA, 1:0.25 HA-MA, 1:0.5 HA-MA, 1:1 HA-MA, 1:2 HA-MA, and 1:3.5 HA-MA ratio. Nuclear Magnetic Resonance (NMR) was used to verify reaction completion and calculate degree of methacrylation. It was seen that as the ratio of methacrylic acid to hyaluronic acid increased so did the degree of methacrylation until a peak of 68% degree of methacrylation for the 1:2 HA-MA ratio was reached. Encapsulation was performed via the air droplet method in which 1 weight percent of the HA-MA ratio was mixed with 4 weight percent of Chondroitin Sulfate A for stability purposes (CSA). Microcapsule membranes and interiors were characterized using Scanning Electron Microscopy (SEM). Photos exhibited thickening of the capsule membrane for higher 1:1 and 1:2 HA-MA ratios. This research set ups future experiments in determining how degree of methacrylation effects capsule compression strength. Moreover, having this data for all the tested ratios helps to narrow down the parameter choices for other future experiments, besides the above-mentioned strength testing, that further optimize the polyelectrolyte complex microcapsules to enhance cell growth, handleability, and integration.

638050 Factors That Influence Successful Ettmp-PEGDA Thiol-Ene Hydrogel Synthesis

Abstract:

Ethoxylated trimethylolpropane tri-3-mercaptopropionate (ETTMP) and poly(ethylene glycol) diacrylate (PEGDA)-based hydrogels are a promising candidate for injectable, localized drug delivery due to the potential for non-swelling behavior and controllable drug release properties. However, we find that existing literature synthesis methods suffer from slow gelation times or a lack of reproducible procedures. In this work, we present a repeatable synthesis procedure, along with a sensitivity analysis of the major variables that impact successful synthesis. Our methods allow the hydrogels to be reliably synthesized in a polymer concentration range of 15 to 90 wt.%, resulting in gelation times between 2 min and 1 h with equilibrium moduli between 3.5 and 190 kPa. Furthermore, base-catalyzed conjugate Michael curing led to 16.35 min and 1.67 min gelation times for buffer pH values of 6.5 and 8, respectively. Additionally, we show that ETTMP purification directly controls the concentration of mercaptopropionic acid that inhibits the Michael addition. Lastly, we find that a stoichiometric mixture of ETTMP and PEGDA freezes at -23.6 °C and can be stored at -20 °C for 2 months without crosslinking, as opposed to 2 days at room temperature, making storage and shipment of mixtures possible.

638155 Investigating the Effect of Penetrant Identity on Micelle Formation and Release Time from Organogels.

Abstract:

Previous research has demonstrated that sodium dioctyl sulfosuccinate (AOT) forms a reverse micelle in mineral oil and diffuses as nanocarrier molecules in an organogels whilst oleic acid releases as single molecules, which is fundamental to our understanding of nanostructure diffusion relationships and applications such as drug delivery. These observations have shown promise for further exploration on the potential for other organic molecules to form reverse micelles in gels. The current study investigates whether two surfactants, Solsperse 3000 and Solsperse 9000, would form reverse micelles, and how much time it would take for the full release in different concentrations of styrene ethylene butylene styrene (SEBS) polymer (10wt% - 40wt%) using the Fourier transform infrared spectroscopy-based method for tracking diffusion. The average diffusivities of Solsperse 3000 and 9000 were measured to be 9.87E-09 cm^2/s and 9.96E-09 cm^2/s respectively, over a full release time of approximately 300 hours which is similar to the release time of AOT but is higher than the AOT calculated value of 4.61E-09 cm^2/s by a factor of two. Additionally, the diffusivities of the Solsperse molecules were compared to oleic acid's 6.21E-08 cm^2/s which showed a larger difference by a factor of approximately ten. Our results show that suggest that the Solsperse 3000 and Solsperse 9000 were similar molecules as confirmed by the H-NMR. Solsperse 3000 and 9000 did not form reverse micelles as confirmed of the structural information from dynamic light scattering on the size of the molecule but rather released as a singular molecule. While oleic acid also releases as a single molecule it wasn't found to be affected by the concentration of the polymer whereas Solsperse 3000 and 9000 were. This yielded the additional conclusion that it takes a single releasing molecule of significant size to be affected by changing polymer concentration.

638156 Microstructure Characterization of PEGDA Hydrogels Using TD-NMR

Abstract:

In this work, hydrogel mesh size was evaluated using time-domain nuclear magnetic resonance (TD-NMR), which is a quick and non-destructive technique. The main goal was to evaluate the use of TD-NMR as an alternative method of hydrogel mesh size determination that could yield more microstructural information than compression testing yet be feasible for routine characterization unlike scattering techniques. Poly(ethylene-glycol) diacrylate (PEGDA) was chosen as a model due to its well cited properties in literature and its prevalence in many biomedical devices. The TD-NMR output is characterized by the longitudinal (T_1) and transverse (T_2) relaxation times of protons with a magnetic field after excitation by radiofrequency pulses. In comparison to mechanical analysis, TD-NMR can provide information of fluids at different confinement environments, characterized by surface-solvent interactions, governed by surface chemistry and pore size.

PEGDA hydrogels were synthesized at polymer concentrations from 15% to 50% and molecular weight from 575 to 3400 Da. These were characterized in a Bruker Minispec mq-20 in solution, after cross-linking, and after swelling the gel to equilibrium. A noticeable reduction in T_2 was observed for PEGDA gels upon cross-linking, indicative of the mobility constraint caused by the formation of the hydrogel network. Linear correlations between polymer concentration and the reciprocal of the transverse relaxation rates ($1/T_2$) were observed for all the three states, as expected from theory. Additionally, T_2 distributions were obtained via the inverse Laplace transform, and literature models were used to calculate mesh sizes from the NMR output. The calculated mesh distribution was found to decrease as polymer concentration increased, consistent with a reduction in the pore size as expected. It was hypothesized that heterogeneous networks would have multiple T_2 peaks and broader distributions than more homogeneous ones. This hypothesis was demonstrated by combining hydrogel disks formulations with different mesh sizes in a single TD-NMR test, as displayed multiple and broader T_2 peaks were obtained in this emulation of a heterogeneous network.

The results demonstrate that TD-NMR is capable of detecting water in different confinement environments, and correlated with changes introduced during hydrogel synthesis. These experiments proved that the TD-NMR technique has the potential of determining multiple mesh sizes for hydrogels. Future work will include the development of a different method of inverse Laplace transform of the NMR signal to resolve broad distributions into underlying multiple pore sizes in a hydrogel sample.

638431 Controlling the Synthesis of Colloidal Lead Selenide Quantum Dots

Abstract:

Controlling the Synthesis of Colloidal Lead Selenide Quantum Dots

Henry B. Anderson[1], Audrey O. Darus[1], Tyler D. McCrea[1], Gregory S. Herman[1], Haori Yang[2]

[1]School of Chemical, Biological, and Environmental Engineering, Oregon State University, Corvallis, OR, USA.

[2]School of Nuclear Science and Engineering, Oregon State University, Corvallis, OR, USA.

Room temperature semiconductor detector (RTSD) materials for radiation sensing are of considerable research interest. This is due to their potential to compete with the resolution of the current industry standards for radiation sensing while also decreasing the cost of production and operation. One promising material for developing RTSDs is colloidally suspended lead selenide quantum dots (PbSe QDs). Its many favorable properties include a large exciton Bohr radius of 46 nm, a tunable band gap between 0.28-1.3 eV, and an ability to exhibit multiple exciton generation. In this study, we developed a consistent synthetic protocol to produce lead selenide quantum dots with tris(diethylamino)phosphine and oleic acid ligands and showed a correlation between the particle size and the first exciton absorption peak. Particle size was varied by changing reaction temperature and holding time. Reaction temperatures between 120 ºC and 150 ºC and holding times between 3 minutes and 7 minutes were used. A MARS V microwave was used to heat the reactants and characterization was performed using transmission electron microscopy, UV Vis-NIR spectroscopy, energy dispersive X-ray spectroscopy, and X-ray diffraction. We found that our synthetic protocol is capable of consistently producing QDs with a size ranging from 3-6 nm with good uniformity in shape, expected atomic ratio and the expected clausthalite crystalline structure. By comparing the quantum dots size to the position of their first exciton absorption peak, we were able to develop a strong correlation showing that a red shift in the peak occurs with increasing nanoparticle size. With the development of this synthetic protocol, lead selenide quantum dots can now be implemented into RTSD devices to test their radiation sensing properties, and an investigation of their surface passivation over time by the tris(diethylamino)phosphine and oleic acid ligands can be performed using X-ray photoelectron spectroscopy.

637563 Optimization of the Synthesis of PLA Nanoparticles through Flash Precipitation

Abstract:

For the synthesis of Poly-Lactic Acid(PLA) nanoparticles, a method of nano-emulsion is used. This has often raised issue by taking a long time to create as well as inconsistent particle size and quality. In addition to the time limit, the emulsion method uses water, which can cause swelling of the particles. By using a method of nanoprecipitation, or flash precipitation, these issues are removed. Organic solvent and anti-solvent systems are applied instead which reduce the likelihood of the swelling of particles. By using solvents that are known to dissolve PLA, the anti-solvent can then be interchanged until the most optimal nanoparticle is achieved. This method has higher reproducibility as well as potential for scale up for larger batch sizes. The significance of these particles comes from their wide range of uses, but most notably as a drug-delivery method that can help deliver pharmaceuticals that aid in the fight against various kinds of cancers. The current primary purpose of these particles is to have them fight lung cancers by being delivered via inhaler straight to the lung, rather than through injection. This will then avoid the negative affects often experienced by the solvent that is often used for these therapeutics during treatment.

638489 Investigating Stereochemically-Driven Peptide Assembly Via Molecular Dynamics Simulations

Abstract:

Objectives

Injectable hydrogels that utilize short, self-assembling sequences of peptides mimic neural physiology and provide a delivery mechanism to the brain for neural stem cells. These gels are applied with the goal of promoting tissue repair following damage due to injury, stroke or neurodegenerative disease. We investigated the amino acid sequence lysine (K), tyrosine (Y), phenylalanine (F), isoleucine (I) and leucine (L), (KYFIL). This peptide is one of a class of rapidly assembling pentapeptides for injectable delivery (RAPID) hydrogels developed by the Lampe group at UVA. We observed that racemic mixtures of KYFIL exhibit lower stiffness and distinct morphology compared to the enantiomerically pure peptide hydrogels.

Methods

In order to investigate the mechanisms of stereochemically driven gel assembly, we conducted molecular dynamics simulations through Nanoscale Molecular Dynamics (NAMD) and performed analysis through Visual Molecular Dynamics (VMD). We modeled 64 amidated KYFIL peptides, varying the ratio of L:D-form peptides solvated with explicit water containing physiological concentrations of ions. Under constant pressure and temperature conditions, we followed the trajectory of KYFIL in NAMD for 200 ns.

Results

Qualitative visualization in VMD focused on clusters, subgroups of peptides close together. The simulation's trajectory revealed clusters smaller in size but larger in quantity in racemic mixtures compared to simulations with only pure L or D peptides. Enantiomerically pure L and D peptides formed larger clusters in smaller quantities. We observed that in enantiomeric mixtures L-KYFIL and D-KYFIL cluster together, rather than in enantiomerically pure clusters. Further, hydrogen bonding data suggest that the average number of hydrogen bonds after 200 ns decreases in racemic mixtures. This suggests stereochemically driven cluster assembly.

Conclusions

We conclude that in stereocomplexed peptide hydrogels, enantiomeric ratio plays a role in the assembly mechanism of gels through the size and enantiomeric composition of clusters. This difference in cluster assembly seems to be directly connected to experimental observations of change in morphology of these gels. Further investigating the mechanism behind assembly will reveal more information about RAPID gels and their potential clinical applications in neural tissues.

Acknowledgements

We'd like to thank Professor Kyle Lampe, Professor Rachel Letteri, and the members of each of their groups for their continued support, as well as the Center for Advanced Biomanufacturing at UVA for the opportunity to explore this project through their summer internship program. Finally, we would like to thank the Rivanna Computing Cluster at UVA for their guidance in high performance computing.

638358 Polymer Dominated Phase Behavior of Coacervate Pairings with Charge and Hydrophobicity

Abstract:

The highest level of polymer science is full of materials with many applications but can have detrimental effects to the environment due to their dependability on organic solvents, such as DMF and benzene. For example, in high end industrial adhesives a polymer resin is mixed with a solvent to create a solution with improved spreadability. A way to get around this problem is to use materials which form in an aqueous solution such as those created through polyelectrolyte complexation. Polyelectrolyte complexation occurs as an interaction between oppositely charged macromolecules and is affected by varying the compositions of the solution and the chemistry of the polymers. The interaction between these polymer chains can result in various products ranging from solid polyelectrolyte complexes, a liquid-solid phase separation, to complex coacervation, a liquid-liquid phase separation that produces a polymer dense phase, or even a single-phase solution.

As an alternative mechanism of product formation, we explored the formation of polyelectrolyte complexes using various polymers such as PTMAEMA and PMAA. In our experiments, the phase behavior of these polymer systems was evaluated by changing the stoichiometric ratio of the two charged polyelectrolytes and by changing salt concentration to find the stability of the materials in salt. We were thus able to create phase diagrams to map the region where complex coacervation occurs. In regards to salt resistance the trends allow us to analyze the stability of systems under different salinities by changing the charge ratios as well as relative hydrophobicity of the paired polymer systems. By increasing the charge density of a polymer system, the stability increased. When hydrophobicity of polymer chains, increased stability were also seen albeit on a different scale. Through the knowledge of the phase behavior of these materials, further research will be conducted into applications for these systems such as underwater adhesives as well as coatings and films with knowledge of their physical properties in mind, taking advantage of their aqueous processing.

638259 Assembly of Doped Infrared Plasmonic Nanocrystal Cubes

Abstract:

Faceted metal oxide nanocrystals (NCs) are inorganic materials that may be spectrally tuned via controlled synthesis and interact in the infrared spectrum. This interaction results in Near-Field Enhancement (NFE) at the NC surface that allows for photocatalytic and molecular sensing applications. To increase the effectiveness and efficiency of these applications, this work optimizes monolayered assembly of fluorine-doped indium tin oxide ($F,Sn:In_2O_3$) NCs and couples the array with PbS quantum dots (QD) to enhance photoluminescence decay rate. Prior work ruled out spray coating, drop casting, and dip coating as methods of assembly, as the resulting images were not nicely dispersed arrays, but rather, had rings and clumps. The finalized assembly method, liquid-air interface assembly, proved the most effective method for producing consistent monolayers and bilayers of NCs and QDs: solutions of NCs dispersed in hexane were dropped on to viscous, immiscible ethylene glycol and drained onto a silicone substrate. This process was optimized with the creation of a Teflon well that minimized perturbation of the nanocrystal layer, producing even arrays of NCs and QDs, revealed via electron microscopy imaging. Further results with STEM-EELs imaging revealed intensified NFE in the monolayered and bilayered arrays. Ultimately, the PbS QDs were used as a probe to reveal that plasmon-exciton coupling intensifies the NC's electromagnetic field, increasing the rate at which light emits. This work indicates the ability of NC Near-Field-Enhancement to augment the properties of neighboring materials and implies potential in innovating and enhancing energy harnessing applications. More notably, the NCs operate in the infrared spectrum, and despite infrared light's abundancy, most solar technology operates in the visible light spectrum. Thus, the nanocrystals may prove to be a useful tool to tap into this spectrum, broadening the horizons of energy technology for future innovations.

638308 Solvent-Assisted Reshaping of Multicomponent Colloids

Abstract:

Advances in phase separation techniques and micro- and nanoscale fabrication have led to a surge in the popularity and relevance of anisotropic colloids. Due to their asymmetric properties, these colloids show great promise as advanced materials, including solid surfactants, micromotors, and hydrophobic coatings. To further broaden and enhance the applicability of these systems, researchers need to develop new, highly tunable synthesis and fabrication techniques. We present a novel method utilizing the technique of solvent assisted phase separation (SAPS), which is traditionally used on non-crosslinked, biphasic polymer particles to promote complete phase separation. We discovered that when SAPS is performed on partially cross-linked particles with a kinetically-arrested structure, the particle components rearrange to create a new morphology. This technique enables the creation of particles with sharply contrasting surface structures, and a model system is presented showcasing anisotropic surface roughness. We propose a mechanism for the process, highlighting the dependence on the relative amphiphilicities and interfacial tensions of present phases. A potential application of the model system as a reversible solid surfactant is also showcased, along with a demonstration of morphology tunability.

638440 Surface Oxidation of Tris(diethylaminophosphine) Passivated PbSe Colloidal Quantum Dots

Abstract:

PbSe Quantum Dots (QDs) have a large exciton Bohr radius of 46 nm, a tunable bandgap between 0.28 eV and 1.2 eV and exhibit multiple exciton generation. These properties give PbSe QDs excellent potential as the active sensing materials in room-temperature radiation sensors. Surface oxidation severely impacts device quality by restricting the transport of charge carriers. Nanoparticle materials can have their ligand chemistry optimized to prevent surface oxidation. In this research, two precursors, one containing lead (II) oxide powder, oleic acid (OA), and 1-octadecene, while the other one containing selenium powder and tris(diethylaminophosphine) (TDP), were mixed in a glove box and transferred to a Schlenk line to dissolve. Reactions were processed in a batch microwave reactor. Material batches were synthesized using temperatures between 120 °C – 150 °C and holding times of 3 – 7 minutes. The reaction product was purified by adding a poor solvent and centrifuging. The PbSe QDs were characterized by transmission electron microscopy (TEM), energy-dispersive X-ray spectroscopy (EDS), UV-Vis-NIR spectroscopy, Fourier transform infrared spectroscopy (FTIR), and X-ray photoelectron spectroscopy (XPS). The characterization data showed mostly uniform particles with spherical morphology. Both OA and TDP ligands were identified on the QD surface. A first exciton absorption peak was observed in the NIR region. The Pb:Se atomic ratio was found to be close to the ideal value of 1:1. A comparison of qualitative XPS data obtained from fresh and four-day-old material showed no signs of oxidation, which could be most easily observed in the Se 3d spectra. By combining OA and TDP as the passivating ligands, the characterization results demonstrated that the QD surface was successfully passivated.

637735 Hierarchical Assembly of Nanoparticles Onto Electrochemical Electrodes

Abstract:

Electrochemical electrodes depend on the fast transport of ions to and from their surface to interact with the active materials. A hierarchical pore size distribution on the electrode is known to improve the performance of high-rate applications such as supercapacitors, but work to date has relied on polydisperse nanoparticle building blocks that are mixed with additives such as polymeric binders and carbon black. The polydisperse nanoparticles result in uncontrolled pore structure, and the additives complicate the ionic transport and reactive interface. This project aims to solve these challenges by creating a hierarchically structured porous electrode from monodisperse nanoparticles deposited onto a substrate without the use of additives. In this work, monodisperse copper sulfide nanoparticles were assembled onto a metal substrate through electrophoretic deposition (EPD). A setup for the EPD electrodes was designed and optimized, and the ideal type of nanoparticle, type of ligand, method of ligand attachment, and solution concentration for EPD were determined. The Keyence VK-X260 Laser-Scanning Profilometer was used to measure the film thickness and uniformity. Using this data, the average film density and porosity was calculated for each sample. Preliminary results have shown that porosity can be controlled, that film uniformity decreases at high voltages, and that there exists a crossover in film thickness between applied voltages of 400 V and 650 V.

638118 Stability, Structure, and Morphology of Dicarboxylic Acid Particles with Polymer Excipients Synthesized VIA Monodisperse Droplet Evaporation

Abstract:

In the pharmaceutical industry, properties such as structure and stability are very important. Many new drugs show low water solubility/bioavailability (the amount of a drug absorbed by the body). Altering a particle's structure is one way to solve this issue (Van Zee et al., 2020). Particles can have several structures: crystalline (ordered arrangement of molecules), amorphous (random ordering of molecules), or semi-crystalline (mix between crystalline and amorphous). Amorphous particles tend to exhibit greater water solubility but may crystallize over time. Conversely, crystalline particles typically are less water soluble but are more stable. Many compounds have several crystal structures depending on formation conditions. Each of these unique crystal structures is called a polymorph. The properties of an active pharmaceutical ingredient (API) can vary between polymorphs, making them of great importance to the pharmaceutical industry. Particles contain two components: an API and a polymer excipient. Excipients are inactive ingredients that influence structure, stability, and other properties. Moreover, the ratio of API to excipient has an impact on the particles' structure: at very low ratios, stable amorphous particles form; at higher ratios, amorphous particles form that partially crystallize over time; and at the highest ratios, partially crystalline particles form (Vostal, 2018).

This work focuses on two main areas: (1) creating initially amorphous particles and quantifying the amount of recrystallization over time and (2) forming partially crystalline particles and quantifying the amounts of each polymorph present. Particles will be made using adipic acid or succinic acid as the model APIs with poly(vinylpyrrolidone) (PVP) as the excipient. These compounds are chosen since PVP is a common excipient in many pharmaceuticals and dicarboxylic acids resemble many APIs. The Vibrating Orifice Aerosol Generator (VOAG) is used to form the particles in this study, which works like a spray drier, but at ambient conditions eliminating the effects of temperature. The particles are monitored using powder x-ray diffraction (pXRD) and scanning electron microscopy (SEM). Standards of known crystallinity are made using a spray dryer to quantify the amount of crystallization that occurs or the amount of each polymorph present.

Van Zee, N. J., M. A. Hillmyer, and T. P. Lodge, "Role of Polymer Excipients in the Kinetic Stabilization of Drug-Rich Nanoparticles." *ACS Applied Bio Materials*, 3, no. 10, (2020): 7243-7254. https://doi.org/10.1021/acsabm.0c01173.

Vostal, K. "Stability of Amorphous Particles Formed Via Monodisperse Droplet Evaporation." 2018.

637583 Influence of Crystal Size on Crystallization and Structure of Second Component in Diblock Copolymer Films

Abstract:

The use of copolymers to combine the attributes of multiple monomer components have been extensively studied for fields including energy and biomedical applications. The homopolymer polycaprolactone (PCL) displays biocompatibility, hydrophobicity, and the ability to hydrolytically degrade at its ester bond. Polyethylene-oxide (PEO), a hydrophilic biocompatible polymer with well defined mechanical properties displays rapid degradation when exposed to water. These two components may be bonded to form a diblock copolymer, PEO-b-PCL, which combines the catalytic breakdown of PCL along with the strength and water degradation of PEO, the latter of which is often valued in medical applications. Through the manipulation of these two blocks, the resulting material properties may be tailored for a specific use, such as coronary stent coatings or vascular grafts. A better understanding of the material's phase behavior is necessary to properly tune the copolymer, and as such, this study aims to determine the effects of varying solvents, copolymer molecular weights, additive molecular weights, and polymer concentrations on the relative crystallization of each block. Sample formulation employed the solvents toluene and ethyl acetate, four separate molecular weights of the copolymer PEO-b-PCL, and three different molecular weights of additive homopolymer PEO. Fourier Transform Infrared Spectroscopy (FT-IR) was used to analyze the relative crystallization of PEO to PCL while Polarized Optical Microscopy (POM) permitted the study of film morphology and the verification of crystal order. Analysis of the resultant data revealed the solvent interactions exhibited by samples dissolved in ethyl acetate promoted greater relative PEO crystallization. With regards to PEO additives, low molecular weight PEO, roughly 400 g/mol, decreased the crystalline ratio as a result of its liquid-like state altering the order in which each block crystallizes. Higher molecular weight PEO (both 5400 g/mol and 36500 g/mol) added to solutions displayed consistent improvements in relative crystallization across all concentrations. Additionally, multiple concentration saturation points were discovered which provided insight into the effective maximization of crystal ratio for both copolymer in solution and additive in copolymer samples.

638703 Modification of Low Molecular Weight Chitosan with Glycidyl Methacrylate

Abstract:

Introduction

Chitosan is a derivative of chitin, a polysaccharide sourced from shrimp and crabs, among other organisms. A biocompatible and antimicrobial material, chitosan has been used as a hydrogel for 3D-printed scaffolds with medical applications. Recent successes with 3D-printing chitosan include the polymer in mixtures with other hydrogels, such as alginate and gelatin. This investigation targets low molecular weight chitosan, generated by controlled depolymerization, and modified with glycidyl methacrylate (GMA) side chains as the base hydrogel for 3D printing.

Materials & Methods

High molecular weight chitosan was dissolved in 1.5 wt% acetic acid to form a 1.5 wt % solution. Aliquots of this solution were reacted with 0.2M $NaNO_2$. The depolymerization reaction stoichiometry indicates that 1 mole of glycosidic bonds is cleaved per mole of HONO consumed (Allan, 1995). A range of low molecular weight (LMW) chitosan solutions were formed: targeting molecular masses of 10, 20, 30, and 40 kDa.

Glycidyl methacrylate was added to each LMW chitosan solution with a 3.8:1 molar ratio of GMA per glucosamine subunit. Samples were precipitated with ammonia, centrifuged, and washed four times with water before freezing and lyophilization. All solutions will be characterized as 1.5 wt% GMA-chitosan dissolved in 1.5 wt% acetic acid.

Printing and material characterization

0.1% lithium phenyl-2,4,6-trimethylbenzoylphosphinate (LAP) was added to GMA-chitosan solutions prior to exposure with 405nm light for 8 minutes. An Anycubic Photon was used for 3D printing. For tensile testing, GMA-chitosan solutions were injected into a mold and exposed to a 405nm LED.

Results & Discussion

Early tests with the low molecular weight GMA-chitosan have not produced a gel sturdy enough to be removed from a mold or to withstand the shear stress required for stereolithographic (SLA) printing. Presently, the dissolution of the GMA-chitosan with a molar reaction ratio of 3.8:1 has been limited. Future tests with the LMW GMA-chitosan will be run at higher concentrations and lower degrees of methacrylation. Upon discovery of a preferred reaction ratio and concentration, tensile strength of the gel and rheological analysis of the solution before gelation will be reported, along with material characterization.

Allan, G. G., Peyron, M. (1995). Molecular weight manipulation of chitosan II: prediction and control of extent of depolymerization of nitrous acid. *Carbohydrate Research 277.*

638416 A Radical Library: Cataloging Radiation Yield of Acrylates Undergoing Electron-Beam Polymerization

Abstract:

Electron-beam (EB) polymerization provides a fast, environmentally friendly method to create thin films and coatings for commercial applications. Expanding EB technology to new applications requires a better understanding of how starting materials (i.e., monomers) form free radicals under the EB. Free radical formation dictates the ability for the monomer to polymerize and the properties that final polymer exhibits. One measure of this attribute is the primary radical chemical radiation yield, $G(R^{\bullet})$, defined as the number of primary radicals formed per 100 electron volts (eV) delivered. The monomers chosen in this study belong to the acrylate family, as this family is industrially prevalent and also exhibits a range of properties, such as number and type of labile bonds, that can be used to understand the influence of these properties on free-radical formation. The concentration of primary radicals was determined based on the absorbance of DPPH, a free-radical inhibitor, at 520 nm after EB irradiation of an acrylate monomer at a given dose (eV) and line speed (ft/s). By plotting changes in DPPH concentration over different EB exposure times through UV-Vis spectroscopic analysis, the rate of primary radical formation was calculated and used to estimate the $G(R^{\bullet})$ for each monomer in the study. By developing this monomer library, we can relate $G(R^{\bullet})$ to chemical structure and ultimately polymer properties, providing guiding principles for materials synthesis, selection, and processing, rather than trial and error, as is the current industrial practice. This monomer library can be reliably expanded using the methods developed in this study to further the potential uses for EB polymerization in the industrial field and academia.

638447 Improving Fabrication and Analyzing Gas Separation Behaviors of Polyamide Membranes Made Using Interfacial Polymerization of M-Phenylenediamine and P-Phenylenediamine.

Abstract:

Polyamide (PA) membranes made from the interfacial polymerization (IP) of m-phenylenediamine (MPD) with trimesoyl chloride (TMC) are popular for applications such as nanofiltration and water desalination, with some research in CO_2 capture. Despite the conventional fabrication of those membranes involving MPD, little research has gone into the fabrication and testing of one of its isomers: P-phenylenediamine (PPD). PPD has exhibited similar performance when used instead of MPD, typically having slightly lower rejection and higher permeation in both reverse osmosis (RO) and gas separations. Due to the para-orientation, PPD has higher reactivity than MPD during polymerization. This increase in reactivity usually does not affect the fabrication process; however, in studies where larger acyl chlorides were used to enhance flexibility and improve flux, PPD-fabricated membranes contained fewer defects. This resulted in a significantly better reported rejection than in MPD-fabricated membranes. During fabrication, the PPD solutions turned orange as time progressed, indicating the PPD was decomposing in the light. After about a day, the PPD-fabricated membranes and their soaking solution change color to a dark purple/black color, indicating the oxidation of the PPD monomer. The membranes themselves also changed color from blue, to purple, indicating this affected the PA as well. Due to its noticeable reaction in light and oxygen, PPD solutions and membranes were then stored in dark, air-tight environments. Because the PPD monomer can react with the environment and contaminate the membrane, future PPD-fabricated membranes were stirred in a vial of DI water and a magnetic stir bar on high intensity for 5 minutes before storing to encourage the leaching and separation of excess monomer from the membrane after fabrication. The permeance of CO2 in the PPD-membranes proved to be close to MPD-fabricated membranes. Since the MPD-fabricated membranes exhibited no color change while in solution, the procedure of the MPD-fabricated membranes did not change; however, excess MPD on fabricated membranes and reactions with the surrounding environment that do not change the color of the solution could still affect the performance of the membranes.

636653 Characterizing Transport of Crosslinked PEGDA Membranes to Carboxylate Ions of Varying Chain Lengths

Abstract:

The need for renewable energy sources has led to research in numerous technologies including photoelectrochemical CO_2 reduction cells (PEC-CRC) which utilize sunlight and water to reduce carbon dioxide into carboxylates and alcohols that can be used as feedstock chemicals or fuels. One difficulty facing CO_2 reduction cells is the design of new ion exchange membrane (IEM) that minimize permeation of electrochemical reduction products formed (carboxylates and alcohols) while maintaining ionic conductivity (ion transport). Previously, the transport of the carboxylate product acetate has been investigated for IEMs composed of poly(ethylene glycol) diacrylate (PEGDA, n = 13), poly(ethylene glycol) methacrylate (PEGMA, n = 9), and 2-acrylamido-2-methyl-1-propanesulfonic acid (AMPS) but the transport behavior of other carboxylates of varying chain lengths (e.g. formate, propionate, butanoate) has not been previously examined. To address this knowledge gap and further our understanding of how carboxylate chain length and membrane chemistry impact carboxylate permeation behavior in IEMs, four IEMs of varied composition were synthesized (100%PEGDA, 68%PEGDA/32%PEGMA, 68%PEGDA/16%PEGMA/16%AMPS, 68%PEGDA/32%AMPS, all mol%) and characterized for their physiochemical and transport behavior. Overall, as the carboxylate chain length increases, the membrane permeability (P) to each carboxylate decreases across the four membrane compositions. Membranes with smaller crosslinker (PEGDA) content resulted in larger water uptake and presumably larger mesh size resulting in the higher observed permeability. To fully characterize the transport behavior, measurement of membrane solubility (S) to these solutes is ongoing, and the diffusivity (D) will be determined through the solution-diffusion model (P = S*D).

637709 Optimization of Lignin Precipitation with Functional Group Control for Use in Bio-Based Polyurethane Foams

Abstract:

Lignin is an organic polymer that can be precipitated from black liquor, a byproduct of the pulp and paper industry. Lignin has the potential to replace petroleum-derived polyols in polymers, such as polyurethanes. In order to cross-link the polymers, a reaction between an alcohol or polyol, and an isocyanate must take place. The substitution of the lignin polymer, which is a naturally occurring polyol, for the petroleum derived polyol will lead to an improvement in the sustainability of these products. In this project, 3-mercapto-1,2-propanediol was reacted with black liquor during the lignin precipitation process in order to increase the proportion of hydroxyl groups, thereby increasing lignin reactivity and integration into the polyurethane network. Previous attempts to increase hydroxyl content through demethylation using the catalyst dodecanethiol were unsuccessful; however, evidence for incorporation of the alkane chain into the lignin structure was found. This suggested the new approach to incorporate hydroxyls into the lignin polymer using a short chain mercapto-alcohol. Using previously optimized precipitation conditions of a modified Ligno-Boost procedure, the resulting lignin with an increased hydroxyl content were used to create a broad range of sturdy foams. The hydroxyl and sulfur content of the recovered lignin was determined using XPS and confirmed that the expected reaction took place. The control lignin product and reacted lignin product were mixed in various proportions with a synthetic polyol (glycerol propoxylate) and reacted with polymethylene polyphenyl isocyanate to create bio-based foams. Future experiments will evaluate the resulting foams for flexibility, tensile strength, and impact resistance in order to determine how much lignin can be added while maintaining the mechanical properties of petroleum-based polyurethane foams. Increased lignin hydroxyl content is expected to result in more rigid foams due to the increased capacity for crosslinking as well as darker in color due to the presence of lignin. These tests will show how these lignin-based foams can act similarly or better than existing products in industry such as cushions, packaging, furniture, and fiber applications.

638679 Analysis of Glass Transition Temperature Depression of Polyvinylchloride Using Nucleophilic Substitution

Abstract:

Polyvinyl chloride (PVC) is one of the most ubiquitous plastics used in our society. Plasticizers are typically added to PVC to increase its flexibility by lowering the glass transition temperature (). PVC plasticization is most commonly performed using external plasticizers, which are small molecules (*e.g.* phthalates) that physically mixed into the PVC. Leaching of external plasticizers from PVC can degrade the properties of PVC and is a potential health risk.

Internal plasticization of PVC involves altering the chemical structure of PVC to lower the T_g. A significant benefit of internal plasticization is that leaching is typically not observed with internally plasticized PVC. This project focused on internal plasticization of PVC by substituting thiols onto the PVC chain. The effect between the degree of substitution of different thiols on PVC and of modified PVC samples was analyzed. Three thiols were substituted onto the PVC chain: 1-dodecanethiol, 2-ethyl-hexyl-thiol and 1-adamantanethiol. The amount of thiol added was varied to create a variety of degrees of substitution (DS) of PVC. Thermal data was collected using a differential scanning calorimeter (DSC). The thermal data was sampled for modified with thiols at different degrees of substitution to effectively compare the between degrees of substitution for different thiols.

638107 Predicting the Mechanical Properties of Triblock Copolymer Gels

Abstract:

The goal of this project is to create an algorithm that predicts the stress-extension behavior of styrenic triblock copolymer gels. These gels are used in a variety of different applications ranging from model surgery and ballistics gels to bike seat cushions. It is crucial that the gel used for a given application exhibits the desired mechanical behavior. The gels we study are composed of styrenic ABA triblock copolymers, such as poly[styrene-*b*-(ethylene-*co*-butylene)-*b*-styrene] (SEBS), and aliphatic mineral oil. The copolymers form a physically-crosslinked network which explains the gel's mechanical robustness. To characterize the mechanical behavior of these gels, we perform quasi-static tensile tests. The results of which are nonlinear, elastic stress-extension curves. We fit this data using the slip-tube network model, which serves as the structure of our algorithm. The algorithm allows us to analyze trends in experimental moduli values as functions of our input parameters (copolymer concentration, block fractions, and molecular weight). We compare these experimental trends to the theoretical equations that characterize the moduli, and we can determine the residual between them. The results of this are moduli correction factors, as well as the recognition that the copolymer bridging fraction is not constant, but changes with copolymer concentration. These rectifications to the slip-tube network model allow us to bridge the gap between experimental data and theory and as a result make more accurate predictions of stress-extension behavior for any formulation of styrenic ABA triblock copolymer gel.

638329 Thermal Analysis of Sustainable Polylactic Acid Foam Development

Abstract:

Polymer foams are an important class of materials, with applications in automotives, aerospace, leisure, insulation, and packaging. They are a more sustainable option compared to regular polymeric materials since they are less dense and use less material. These foams can be made more sustainable by using a bio-based polymer as the basis for the foam. The thermal characteristics of the necessary components for creating sustainable bio-based polymer foams were investigated. The base polymer that is used is polylactic acid (PLA). Different foaming formulations were made using exothermic, nitrogen-based chemical foaming agents (CFA). The CFA that were used were Axeleron and Tracel brand solid pellets, and Avient brand liquid CFA. Additionally, polymer crosslinking agents such as dicumyl peroxide (DCP) and tryallylisocyanurate (TAIC) were added to some of the formulations. These components were compounded with PLA via two solid-state processes: cryogenic milling and solid-state shear pulverization (SSSP). Thermogravimetric Analysis (TGA) and Differential Scanning Calorimetry (DSC) were performed on pure, dried PLA pellets, each of the CFAs, and on TAIC and DCP to understand individual degradation temperatures and the foaming activation temperatures for each of the CFA. TGA and DSC were also conducted on the compounded formulations. Comparing the thermal properties of the foaming material formulations with individual material ultimately provides insight on the thermal tendencies of PLA foaming.

638470 Converting Birch Bark Extracts into Bio-Based Thermosets

Abstract:

Birch tree barks are regarded as waste in the pulp and papermaking industry and used as fuel. However, this material presents a source that contains many bio-based chemicals suitable for applications ranging from pharmaceuticals, plastics and composites, coatings, and antifeedants. Among the mixture of bio-derived chemicals in birch barks, triterpenoids, such as betulin, betulinic acid, and lupeol, can be present up to 30% weight of dry bark mass. They are highly valued for their anti-tumor, HIV, and inflammatory responses. In our presented work, triterpenoid mixtures were extracted through a Soxhlet extractor using the barks from locally sourced river birch trees (Betula nigra) with an average yield of 10.6% (dry bark mass). The extracted materials were characterized using the Nuclear Magnetic Resonance (NMR), Advanced Polymer Chromatography (APC), High-Performance Liquid Chromatography (HPLC), and hydroxyl number titration to assess the identity, average molecular weight, triterpenoid content, and the number of reactive sites, respectively. The extracts have been used to synthesize bio-based polymers with promising thermal and mechanical properties using minimal processing steps. Birch bark extract naturally contains many potential reactive sites and thus making it advantageous for synthesizing polymers without requiring multiple purification steps. We demonstrate the potentials for increasing the utility of birch bark, contributing to sustainability challenges in materials science and engineering.

637641 Thermomechanical Characterization of Recyclable Epoxies By the Diels-Alder Reaction

Abstract:

Epoxy products are typically non-recyclable and difficult to remove. Synthesis and characterization of recyclable epoxies crosslinked by a Diels-Alder (DA) reaction were investigated. The recyclable epoxies were synthesized between maleimide and furan functional group to form cyclo-adducts (e.g., polyetheramines and epoxides precursors). Different architecture (e.g., 4-arm or 6-arm) and stoichiometric ratios of furan to maleimide groups were examined to determine the optimum functionality of the recyclable epoxy. The epoxies demonstrated calorimetric glass transition temperatures (T_g) in the range 0 to 30°C depending on molecular composition, and a high-temperature (~120°C) endotherm corresponding to the retro-DA (rDA) reaction was observed. In addition to the low-temperature glassy phase, rheometry detected a rubbery plateau at moderate temperature followed by a rapid decrease in modulus indicative of liquid behavior. This high temperature (above ~140°C) decrease is a consequence of the rDA reaction depolymerizing the network. Plasmonic titanium nitride (TiN) nanoparticles have significant visible light absorption and were incorporated into the recyclable epoxies for targeted and instantaneous triggering of the rDA reaction by a high-intensity light source. The TiN filler (<1.0wt%) slightly increased the modulus of the epoxies and permitted localized heating. The epoxy was further tested by looking at its reusability. Two glass slides were attached using the reversible epoxy; and the sample was then heated using a high-intensity light source. This light was converted to heat in the epoxy by the TiN nanoparticles, causing the epoxy to soften and the slides to separate under stress. The rDA reaction was evident after separation, as epoxy remained on both glass slides. This showed that the epoxy did not simply succumb to adhesive failure. After the slides separated, they were capable of reattachment by exposing the sample to the light source and pressing the slides together. These healed slides were then able to undergo the same stress as before the photothermal test without failing.

638549 Evaluation of Small Molecule Dopants in Self-Healing Conductive Polymer Complex PANI:Paampsa.

Abstract:

Self-healing is a highly beneficial, and desirable, attribute of biological systems that allow them to autonomously repair surfaces in the event of mechanical damage. Recently, flexible/stretchable electronics and wearable sensor researchers have been motivated to imitate this behavior in innovative devices with unique and unprecedented electro-mechanical properties. Conductive self-healing polymers have been utilized to improve a range of electronic and optoelectronics devices, including wearable sensors and displays for healthcare,[1] virtual reality, and soft robots, among many others.[2] For a polymer to be used in wearable sensors, it must possess comparable properties of skin in terms of stretchability (maximum strain), elasticity (Young's modulus), and self-healing—being able to regain both mechanical and electronic properties with a high efficiency. Self-healing properties are necessary to be resistant to internal and/or external mechanical damages caused by wear and tear, cuts, and rips.[8] Additionally, the polymer sensors must have excellent mechanical properties with a high conductivity for sensitive measurements that are reliable and repeatable.

To this effect, our laboratory has previously developed a conductive polyaniline (PANI)-based polymer complex with high stretchability, elasticity comparable to human skin, and autonomous, repeatable self-healing properties.[3] However, further improvement of conductive and electronic sensitivity is desirable to accurately measure slight changes in movement or pressure. To accomplish this, alternative small molecule dopants (SMD) have been added in replacement of the previous dopant, phytic acid. The investigated polymer composite is composed of PANI, a small molecule dopant, and a polyelectrolyte [poly(2-acrylamido-2-methylpropane sulfonic acid) (PAAMPSA)]. The SMDs being evaluated for this polymer are 1) trimethyl 1,3,5-benzenetricarboxylate, 2) 1,2,4-benzenetricarboxylic acid, 3) trimesic acid, 4) formylbenzene which will be referred to as SMD1, SMD2, SMD3, and SMD4, respectively, for this experiment.

The implications from this study will seek out to ameliorate the performance, accuracy, and robustness of polymer sensors in wearable applications. The ability to monitor subtle changes in physiological signals using flexible electrical devices can be valuable for detecting disease through continuous comfortable monitoring as well as be applied for human pulse detection, speech recognition, and electrical artificial skin. The ideal polymer would have high conductivity, elasticity, self-healing capabilities, tensile strength, omnidirectional stretching, and would be scalable at a low cost.

[1] E. K. Wujcik, et al. IEEE Sens. J. 2013, 13, 3430;

[2] C. X. Zhan, et al. Journal of Materials Chemistry C 2017, 5, 1569;

[3] Y. Lu, Z. Liu, H, et al, ACS Appl Mater Interfaces 2019, 11, 20453.

638379 Synthesis of Liquid Crystal Elastomers Functionalized with an Amino Cinnamate

Abstract:

Liquid crystal elastomers (LCEs) are materials that combine the property of molecular order with rubber elasticity. When the liquid crystal (LC) units are well aligned, these materials experience significant, reversible shape change under exposure to heat or light. Current methods for alignment remain limited. A method for obtaining LCEs with little restriction on alignment will enable advancement of the field of programmable polymer materials. The approach of this project is to develop a technique for LCE alignment by incorporating UV light responsive cinnamate derivatives into the LCE. Over the past summer, cinnamates functionalized with free amines and amine salts were tested for incorporation in LCEs using the aza-Michael addition reaction. FTIR kinetics studies were performed on these reactions to determine the reaction rates of the amines. From the results, the amine salts did not incorporate while the free amines successfully underwent the reaction.

638605 Accessing Complex Nanoscale Packing with Aba' Triblock Copolymers

Abstract:

Formed by combining chemically distinct polymer blocks, block copolymers are used across several industries owing to their ability to self-assemble at the nanoscale. Of particular interest in recent years has been the discovery that these ostensibly simple materials can order into micelles arranged on large, complex Frank–Kasper (FK) lattices. Frank–Kasper (FK) phase formation in diblock copolymer melts is largely based on three parameters: block volume fraction, molecular weight, and conformational asymmetry. Less explored, however, is how their formation can be driven in triblock copolymers. This research studies this question with a particular focus on ABA′ triblock copolymers with variable core (A) block lengths. We hypothesize that an asymmetry in the core block lengths will open up a FK phase window. Polymers will be synthesized by ruthenium-mediated ring-opening metathesis polymerization (ROMP). Volume fraction and core block length asymmetry will be varied to yield symmetric and asymmetric triblock copolymers. Polymers will then be characterized by proton nuclear magnetic resonance spectroscopy (^{1}H NMR) and size exclusion chromatography (SEC). Following, small angle X-ray scattering (SAXS) will be used to assess polymer nanostructure and determine order–disorder and order–order transitions.

638688 Fabrication and Characterization of Lignin-Based Soft Composites

Abstract:

Lignin, a biopolymer and byproduct of paper making, has recently garnered attention as a green alternative to petroleum-based precursors in a wide array of applications. Due to lignin's attractive properties, lignin-containing hydrogels have potential to replace materials traditionally leveraged in applications such as tissue engineering, membrane-based water purification, and drug delivery, to name a few. However, the lignin traditionally utilized is oftentimes unpurified, leading to lignins with high dispersity in molecular weight (MW), making it difficult to accurately characterize material properties. In this study, fractionated lignins of prescribed MW are used in the synthesis of soft composites. First, membranes with copolymers of poly(vinyl alcohol) (PVA) and lignin, of two prescribed MW and concentrations as high as 40 mass%, were fabricated using glutaraldehyde (GA) as a crosslinker. Both the mechanical and transport properties of these membranes were characterized. The Young's modulus, water uptake, and number of monomers between crosslinks of the membranes indicated the formation of a tighter network with increasing lignin MW, as well as increasing lignin and GA concentration. Next, we attempted to fabricate composites containing a more biofriendly polymer, poly(2-hydroxyethyl methacrylate) (pHEMA). Unlike the lignin–PVA system, we were unable to synthesize robust, free-standing membranes, though we found that the properties of membranes containing pHEMA as the copolymer were more sensitive to the pH of the environment during membrane fabrication.

638234 Structure-Property Relationships of Bio-Based Thermosets from Birch Bark

Abstract:

Plastic pollution is a pressing issue that threatens ecosystems around the world. Many commonly-used plastics have long lifetimes and are often discarded in the environment. Additionally, plastic manufacturing taxes the environment and diminishes natural resources. Thus, developing bio-based alternatives to petroleum-derived plastics is a promising approach towards lessening the adverse environmental impacts of plastic use. This study focuses on the valorization of betulin, a pentacyclic triterpenol, which can be extracted in high quantities from birch bark. Previously, we incorporated betulin into a series of polyester networks with varying bio-based diacids. The focus of the current study is to develop and characterize the properties of betulin-based thermosets synthesized with two diacid co-monomers in varied ratios and glycerol as the crosslinker. We use dynamic mechanical analysis to examine thermomechanical properties and thermogravimetric analysis to investigate degradative temperatures. Our results show that the storage moduli at room temperature range from 44.6 MPa to 199 MPa depending on the composition and exhibit thermal stability above 296°C in both inert and oxidative environments. This work is part of a larger study to develop and evaluate bio-based plastics at all stages of life, from monomer extraction, to polymer synthesis, to end of life. These results demonstrate that thermosets with promising and tunable properties can be synthesized using betulin and other co-monomers.

638318 Exploration of Curing Conditions for Producing a Soybean Oil-Based Thermosetting Polymer for Potential Use As a Renewable Matrix for Composites

Abstract:

Due to increased environmental concerns, there is interest in exploring bio-based monomers and polymers to produce thermosetting resins and ultimately composites. Our project focuses on thermosetting polymers, meaning they become irreversibly hardened after being cured. Traditionally, thermosetting polymers are produced from petroleum-based chemicals. We are interested in replacing the traditional methods with bio-based alternatives, which are also targeted as possible matrix materials for natural-fiber or natural-particle reinforced composites.

Solid polymers were prepared from a bio-based liquid mixture consisting of acrylated epoxidized soybean oil (AESO) and isobornyl methacrylate (IBM), cured, with a free radical initiator. Previous work from other Bucknell students proved a two step curing sequence of 120 C followed by 160 C to be successful for an 80:20 ratio of AESO to IBM - this produced a solid with mechanical properties appropriate for future composite formation.

A further goal is to optimize curing time and temperature for this system. Minimizing temperature is desirable because it makes the process easier to work with, and potentially makes samples less susceptible to cracking and forming bubbles. The lower temperature is also less likely to damage the plant-based fillers. Combinations of different temperatures and times have been studied for curing AESO and IBM to get an understanding of when curing is complete, and also what minimum temperature can be used to produce a solid polymer. Results from material properties tests were compared to determine how the method of curing affects the physical characteristics. Mechanical analysis included testing the hardness with a durometer and deformation in compression and bending on a universal testing machine.

Modification of the initiator system with promoters, such as Cobalt(II) Naphthenate, is also being explored in an attempt to achieve curing at lower temperatures and time. It is expected that the addition of this promoter will speed up the process and require less heat. The outcome desired from experimenting with different promoters is to achieve a cure at or near room temperature.

638335 Next-Generation High Refractive Index Polymer Thin Films

Abstract:

Conformal optical coatings with high refractive indexes are crucial components in optoelectronic and energy applications. High refractive index polymers (HRIPs)are particularly compelling due to their desirable mechanical properties, such as flexibility and resiliency. However, it is challenging to prepare polymer thin films with high index (>1.8) while retaining optical transparency over visible wavelengths. This project builds on previous work where the refractive index of poly(4-vinylpyridine) (P4VP) was increased to over 2.0 through incorporation of halogen compounds. Here, we examine an alternative polymer derivatization approach and characterize the resulting refractive index, thermal stability, and optical transparency. Using initiated chemical vapor deposition (iCVD), 4-vinylpyridine is deposited onto Si wafer substrates and rapidly polymerized, producing P4VP thin-films with well-controlled thicknesses ranging from 400nm to 1000nm and RMS surface roughnesses less than 1 nm. P4VP films are then chemically derivatized with substituents bearing high molar refractivity in sealed jars at 100°C for 0.5 hours to 30 hours. Film thickness and optical properties are characterized using variable angle spectroscopic ellipsometry coupled with an *in situ* thermal cell. After a 6.5 hour treatment, the derivatized P4VP possesses a refractive index of 2.06, accompanied by a 10% increase in thickness and a thermal decomposition temperature well above 200°C. This is a significant improvement over previous results where charge transfer complexes degraded rapidly in ambient (room temperature) conditions. We further demonstrate the ability to selectively tune the refractive index of P4VP by changing reaction conditions during chemical derivatization, opening up optoelectronic applications for next-generation thermally robust HRIPs.

637690 Incorporating Degradable and Reconfigurable Monomers into Vinyl Thermoplastics and Thermosets

Abstract:

Within the waste management system of plastics, polystyrene (PS) and polyvinyl chloride (PVC) are of the least recycled even though they are thermoplastics, recyclable polymers. This is mainly due to
their chemical and physical properties. PVC produces gaseous hydrochloric acid when heated while PS is
a costly material to recycle. A possible solution to reducing the difficulty in recycling PS and PVC would
be to incorporate reconfigurable comonomers into these vinyl polymers. Doing so could allow the polymers to more easily solvolyze into oligomers or crosslink into vitrimers. Through an SN2 reaction,
Meldrum's acid is reacted with carbon disulfide and 1,4-dibromo-2-butene to synthesize a monomer for
radical ring-opening polymerization. However, after attempting to polymerize the compound using several initiators and at various temperatures, it was determined that the monomer was not reactive under these conditions. A more strained compound was computationally identified for future studies.

638159 Characterizing Styrenic Triblock Copolymers Using Static Light Scattering

Abstract:

Light scattering refers to the process in which light hits a particle and scatters in a predictable pattern. This scattering pattern can be interpreted to understand the characteristics of the scattering particle. For this research, a measurement technique called static light scattering (SLS) was used, and due to the characteristics of the measurement setup and the size of particles in solution exclusively Rayleigh scattering is observed. SLS measures the standardized intensity of the scattering at multiple angles and multiple concentrations. This technique allows us to measure constants such as molecular weight, the A2 virial coefficient, and radius of gyration of the particles. Our focus was on characterizing styrenic block copolymers in order to understand the relationships between their molecular structure and their behavior in solution. The instrument used for these measurements is the BI-200SM Goniometer. Each applicable polymer was dissolved in chloroform, stirred overnight, and filtered prior to SLS analysis. Using the known weight concentrations of the solutions, and by measuring the scattered intensity at multiple angles of each concentration, we are able to organize the data into a Zimm plot. This Zimm plot allows constants related to the behavior of the polymers to be found in the slope of the trendlines of the measured data. Patterns between the known characteristics of these polymers and the newly measured constants allow a deeper understanding of the general solution behavior of these styrenic triblock copolymers. An important constant related to the particle's scattering behavior in solution is the refractive index increment. This increment was shown to be directly proportional to the styrene copolymer concentration. Another important finding was that the scaling of the measured radius of gyration with the measured molecular weight matches the expectation for a polymer in a good solvent. The main conclusion from the measured data is that you can effectively characterize styrenic triblock copolymers using static light scattering methods.

636479 Effects of Chemical Characteristics on the Surface Mechanical Properties of Block Copolymer Micelles at the Air-Water Interface

Abstract:

Every year, over 200,000 individuals in the US suffer from acute respiratory distress syndrome (ARDS). ARDS has a high mortality of ~ 40%, while there is no lung surfactant (LS) drug that is effective in treating this disease. To address this need, our laboratory has recently developed a fully synthetic polymer lung surfactant (PLS) therapeutic (polystyrene-poly(ethylene glycol) (PS-PEG) nano micelles) that can perform the surface tension-lowering function in the lungs of an ARDS patient that are deficient of natural LS. PLS produces near-zero surface tension (under high compression) by formation of insoluble monolayers at the air-water interface. In the present work, in order to understand the effect of polymer chemistry on PLS performance, the surface mechanical and molecular conformational properties of nano micelles formed by various different block copolymers (i.e., block copolymers having different hydrophobic polymer blocks and/or different hydrophilic PEG chain end groups (hydroxy vs. methoxy)) have been investigated by surface tension-area isotherm, ^{1}H NMR spectroscopy, and water contact angle measurements. We found that micelles derived from block copolymers with stronger hydrophobicity, higher glass transition temperatures, and the less hydrophilic methoxy group at the chain end produced greater surface tension-reducing effects, which makes them more suitable for use in ARDS treatment. A complete understanding of how the chemical characteristics of polymers affect their surface mechanical behavior will enable us to design improved PLS candidate materials for applications in ARDS therapy.

638649 Liquid Crystal Coated Fabrics

Abstract:

Cholesteric liquid crystals (LC) have the optical property of reflected color because of their helical structure. Their color changes in response to a temperature change or mechanical stimuli due to changes in the helical pitch. Thus, incorporating such liquid crystals into fabrics is of interest to achieve responsive materials. To achieve responsive materials in this work, we examined liquid crystal fabric coatings. The effect of fabric type, emulsification of the liquid crystal before application, as well as the use of top coats on the responsive properties of the fabric were investigated. Specifically, six fabrics (knit cotton, twill cotton, twill lyocell, knit rayon/spandex blend, woven rayon/spandex blend, and woven nylon) were coated with liquid crystal alone, liquid crystal-polymer emulsion, or a liquid crystal-polymer emulsion with a polyurethane top coating. Image analysis in CIE L*a*b* color space was performed to determine if the color of the liquid crystal was affected by processing. Overall, emulsification and coating of the liquid crystal in a water soluble polymer improved the brightness (increased L-value). Processing (emulsification and application to the fabric) did shift the color compared to the liquid crystal alone as the color difference (ΔE) was 24.0 ± 5.0. Further analysis using UV-vis reflectance spectroscopy and polarized light microscopy are ongoing.

638381 Manipulating Methylamine Deintercalation in 2D Halide Perovskites Via Chain Length

Abstract:

The need for renewable power has driven the search for new spaces in which to implement photovoltaic technologies. One such space is photovoltaic windows. A potentially desirable functionality for photovoltaic windows is the ability to change transparency. To date, halide perovskites have demonstrated such a color change in addition to high photovoltaic performance. Halide perovskites, when exposed to methylamine gas, form a clear phase and return to a dark phase when the gas is driven out. This color change is the result of methylamine intercalation into the halide perovskite material, and the concomitant structural rearrangements, followed by deintercalation. Methylamine intercalation/deintercalation can result in morphological changes which would lead to photovoltaic performance loss. Thus, templating the halide perovskite structure is desirable to improve reversibility. 2D Ruddlesden-Popper phase halide perovskites with the formula A_2PbI_4 are explored and assessed for methylamine intercalation/deintercalation. Specifically, the length of the A-site organic spacer was varied to gain insight into appropriate design rules for these materials. The primary organic spacers studied are phenylalkylamines with alkyl carbon chain lengths of 0, 2, and 4 between the phenyl ring and ammonium head group. With varying chain lengths, there are differing intermolecular interactions between the carbon chains in the dielectric layer of the 2D halide perovskite and thus differing molecular forces. It is found that 2D perovskites with weak intermolecular interactions in this organic dielectric layer retain significant amounts of methylamine and show greater, irreversible, structural changes upon methylamine intercalation/deintercalation. Ultimately a better understanding is gained of how differing A-site organic spaces can template the 2D halide perovskite structure and lead to more reproducible intercalation and deintercalation of MA gas. This provides improved understanding of the complex interactions in 2D halide perovskites and provides a step toward improved color switchable perovskite solar cells.

638427 Fabrication of 2D Textile Patterns Via Electrospinning - Achieving Negative Space Features

Abstract:

Electrospinning is a simple method of producing polymer fibers. Controlling the spatial arrangement of fibers to achieve patterned morphologies has attracted attention for various textile, filtration, and sensor applications. Metal templates such as grid electrode collectors create an "electric field lensing" effect that result in hierarchical self-assembly of the fibers. Overall, we aim to achieve fabrication of 2D textile patterns using electrospinning and such metal templates. Arbitrary garment shapes (e.g. single use mask, apron) were obtained by templating the grid electrode collector with insulating material. The focus of this work is to attain a negative space feature (e.g. holes) during electrospinning using nylon fibers spun from formic acid as a model system. Pattern features, specifically hole size and location, as well as utilizing various insulating methods such as air gaps are being explored. Air gaps and other insulating materials reduce fiber deposition on specified areas of the template. Using this approach, holes at the edge of the pattern (e.g. ear loops of masks, neck piece of an apron) have been achieved. In contrast, holes in the center of a pattern are more difficult to achieve as fibers accumulate over the desired negative space. Use of tape as insulating material over the desired hole produced the best results compared to foam or rubber: the area covered was 88% of the expected area. Foam and rubber decreased the area covered of the desired pattern to 86% and 60%, respectively. Experiments to understand the minimum distance from the center necessary to successfully achieve a hole as well as the smallest hole (relative to the width of the pattern) are currently underway.

638507 Thermochromic Nonwoven Fabrics Via Electrospinning

Abstract:

Cholesteryl ester liquid crystals (LC) are a class of unique soft materials with thermochromic properties arising from their helical structure with a temperature-dependent pitch length. Near the mesophase transition temperature, the liquid crystal first reflects blue light ($\lambda = 450$ nm) at relatively short pitch lengths. As the temperature decreases, the wavelength reflected shifts to the red end of the spectrum ($\lambda = 760$ nm) due to the increase in pitch length. Such materials have been used in thermal mapping and analysis in medical, industrial, and engineering applications. The LC in this study are composed of cholesteryl oleyl carbonate, cholesteryl pelargonate, and cholesteryl benzoate (30:60:10 by weight).. Electrospinning is a useful approach to incorporate such functional additives into nonwoven fibers. In this work, we aim to achieve thermochromic nonwoven fabrics in a single-step, single nozzle electrospinning process using solvent induced phase separation. Specifically, we blend the LC with spinnable polymer solutions (one-phase). During the fiber spinning process, as the solvent evaporates and the polymer concentration increases, phase separation of the liquid crystal will occur. Polycaprolactone (PCL) (10 wt.% which formed uniform fibers alone) was blended with the LC and electrospun using chloroform as the solvent. The LC concentration was varied between 2.5 and 27.5 wt.%. The fibers were analyzed using polarized light microscopy (PLM) on a temperature-controlled stage. At low LC concentrations (below 15 wt%), no thermochromic behavior in the fiber was observed. Increasing the LC concentration above 15 wt.% resulted in non-uniform fiber formation and nonwoven fabrics with an oily feel. Thus, the intermediate LC concentration of 10 wt% PCL 15 wt% LC-3 ideal for achieving uniform fibers and robust fabrics. The resulting fibers and fabrics were thermochromic and demonstrated reversible color change with no significant change in color transition temperatures over at least 4 cycles. Overall, we demonstrate that electrospinning is a promising approach to achieving thermochromic nonwoven fabrics.

637428 UV Curable Polybutadiene-Based Resin for 3D Printable Polymer Electrolyte

Abstract:

Constructing a viable UV-curable gel polymer electrolyte is essential to the development of 3D architected lithium-ion batteries with high energy and power density. Here, we report our progress on the development of a UV-curable polybutadiene-based gel polymer electrolyte with measurable ionic conductivity and mechanical strength. The UV-induced crosslinking of the polybutadiene network is achieved through the formulation of a photo-resin containing a crosslinker (1,4-butanediol diacrylate), photo-initiator (TPO-L), and photo-blocker (SUDAN I), and has been corroborated through the Raman spectra taken from resin samples before and after UV-curing. The mechanical strength of the crosslinked polybutadiene is tested using a G200 nanoindenter to ensure that sufficient strength is achieved by the crosslinked polymers. A promising ionic conductivity on the order of 10^{-6} S/cm is observed via electrochemical impedance spectroscopy (EIS) upon swelling the crosslinked polymer in solutions of 9BBN and organolithium reagent. The resulting polymer electrolyte has wide implications – the future steps include incorporating the electrolyte into a conventional battery system to study the cycling performance and explore the potential to 3D print the polymers for applications in architected electrode systems.

638017 Direct Ink Writing for Sintering of Tungsten Metal Parts

Abstract:

Tungsten (W) metal is known for its ability to withstand high temperatures with a melting point at 3422°C. W materials are also durable, dense, and thermally conductive. Currently, there is no viable method of fabricating W in complex shapes, but this ability would be extremely useful. These complex W parts would be able to be used for nuclear-thermal-propulsion (NTP) engines for space travel, hypersonic aircraft leading edges, and solar sensors that require specifically designed parts.

The purpose of this project is to use direct ink writing (DIW), a type of 3D printing, to fabricate metal-doped tungsten parts. Certain dispersed metal dopants can be used to reduce the sintering temperatures of tungsten, and DIW allows for the W to be printed into complex shapes. Because there is no current way of making complex W shapes, this is a very novel method.

An initial W ink was formulated using tungsten powder, water, a binder, dispersant, and flocculant. The ink is uniform, of ideal consistency, and prints solid figures that hold their shape. The method of printing was studied using a design of experiments. The size of the syringe tip that the W ink flows out of, the speed at which the shape is printed, and the space between each layer when the part is printing were each varied. The parts were then sintered in a furnace at 1800°C and their densities and surface roughness were analyzed to determine how each factor, or combination of factors, affects the quality of the printed part. This allowed us to find the optimal ink to be used to print complex tungsten shapes.

638457 Investigation of the Impact of Equivalent Weight on 3M Proton Exchange Membrane Functionality

Abstract:

Polymer membranes are an integral component of Proton Exchange Membrane (PEM) fuel cells and are therefore integral to advances in energy conversion and storage and a cleaner energy future. Much research has been conducted on the class of membranes known as perfluorosulfonic acid (PFSA) membranes due to their promising performance characteristics. Recently, many new PFSA membranes with variations in structure, composition, and equivalent weight have been synthesized and characterized in hopes of improving efficiency and reducing cost, currently the major barriers to more widespread implementation. As side chain chemistry largely determines membrane functionality, equivalent weight is an important parameter to explore. In this work, the effect of equivalent weight on PFSA membrane properties was studied. Specifically, PFSA membranes created by 3M with equivalent weights of 725, 825, and 1000 were tested to determine conductivity, water sorption, and specific density. Conductivity was tested over a range of temperatures and relative humidity levels simulating potential operating conditions. Water sorption was tested over a range of relative humidity levels to gain an understanding of how relative humidity affects membrane hydration, which is one of the primary factors determining the rate of proton transport in the membrane. In addition, specific density was also tested at these conditions, as density offers insight into membrane hydration and polymer morphology. By studying membranes in this fashion, their performance can be objectively assessed in a controlled environment prior to testing in PEM fuel cells. Membranes with lower equivalent weight exhibited higher conductivity, higher water sorption, and lower specific density. This suggests that for the range of 3M membranes studied, lower equivalent weights, i.e. higher concentrations of sulfonic acid groups, contribute to better membrane performance. Through these findings, this study attempts to further the understanding of the impact of side chains on membrane functionality and guide membrane synthesis research moving forward.

638053 Exploring 3D Printing of Pure Poly(ionic liquids) and Blends with Commercial Resins

Abstract:

Stereolithography (SLA) is an additive manufacturing (3D printing) process where objects are printed layer by layer via photopolymerization and crosslinking using a liquid resin. This project focuses on the integration of poly(ionic liquids) (PILs) into value-added composites with commercial resins, processed via SLA. This work explores the utilization of two commercial photocurable resins, ABS or acrylate-based, interfaced with common vinylimidazolium-based PILs: [C2vim][Tf2N] and [C4vim][Tf2N] to analyze the influence of functionality, ionic content, and composition on the conductivity and mechanical properties of each composite. A wide range of suitable mass ratios (PIL:Resin ~ 90:10 – 0:100), were studied in order to evaluate the structure-property relationships of these 3D printed blends. The SLA 3D-printing conditions were developed and optimized for each composition, utilizing a benchtop SLA-LED printer (λ ~ 405 nm, I· = BAPO), the incorporation and curing of the PIL and ABS or acrylate components was confirmed using FTIR. The physical and mechanical properties were evaluated directly after printing or following a post-print curing step.

637990 Developing a Rapid Prototyping Microfabrication Technology with Conventional Photolithography Techniques

Abstract:

Microfabrication can be accomplished by conventional photolithography techniques or by newer rapid prototyping techniques, which speed up the fabrication process. Most approaches to rapid prototyping involve entirely re-designing the microfabrication process, such as by inkjet printing with fabrication materials and substrates. Many materials and devices, such as those produced in glass or silicon, are best fabricated using conventional photolithography. Our research project will increase the throughput of microfabrication by photolithography by developing a rapid prototyping technique to produce high precision, low cost photomasks. The new technique will utilize the Sony XPERIA Z5 smartphone, which has the highest pixel density (806 ppi) of any smartphone currently available, to produce high resolution photomasks by exposing ortho-lithographic photography film in surface contact with smartphone screen. The project will take part in three phases. Our goal is to develop a custom android app to control photomask geometry, along with a custom designed 3D model for phone and film placement, and to test exposure time and the film development process, lastly we will work on implementing all of these results on the Sony XPERIA Z5. Microfabrication is significant in many ways, and even more so developments in their process, for our research we focused on the significance for NASA and more specifically finding life elsewhere in the solar-system. Life elsewhere in our solar system is most likely to be found in aqueous environments, such as the subsurface oceans of Europa or Enceladus. Microfluidic devices are exceptionally well suited to characterizing chemical biology in aqueous samples, and their small footprint, low energy consumption, and automated operation make them ideal technologies for solar system exploration. The proposed research will provide an enabling technology for the development of new microfluidic devices that will support the search for life in our solar system. Currently in this phase of research our results are more developmental products, then data values. An app to control photomask geometry has successfully been developed and can be used on any android device. The film development process has also been developed and tested on a basic level. Development for the 3D model has gone well but it is still undergoing work to make sure the process is as simplified as possible due to the nature of the work. In these early stages of the research the conclusions look promising, we have been able to develop every step of the process so far without any hitches.

638149 Photopolymerization of Semi-Crystalline Polymers: Fundamental Characterization and Opportunities for Additive Manufacturing

Abstract:

With the growing importance of additive manufacturing (AM), there is a need for new materials development to match the property demands. While extrusion-based AM has been applied to many material classes, the property scope of resins amenable to light-based approaches to AM is limited. One challenge in this area is the development of formulations that polymerize rapidly at low intensities and exhibit targeted, high performance mechanical and physical properties upon curing. Most conventional photopolymerization-based AM resins rely on chemically crosslinked polymer networks to improve cure speed, but these resins tend to be brittle and lack the ability to be thermally reprocessed. To address these challenges, our group has recently developed a new class of semicrystalline thiol-ene polymers that polymerize rapidly in a spatiotemporally controlled manner to form linear polymers that can be melted at reasonable, super-ambient temperatures.

In this work, the goals were to investigate the relationships between polymer structure and properties such as molecular weight, melting temperature, and kinetics of polymerization and crystallization. By implementing off-stoichiometric polymerizations, the molecular weight and thermomechanical behavior of the polymer was controlled. The ability to tune these mechanical properties and the kinetics of polymerization and crystallization is useful in 3D printing materials designed for applications such as cast molding jewelry, dental implants, or other customized molds.

The focus of this work was on the step growth polymerization of the thiol-ene "click" reaction of diallyl terephthalate (DAT) and hexanedithiol (HDT) initiated by TPO. The molecular weight of the polymer after polymerization was studied by GPC. Polymerization kinetics were measured with FTIR, and crystallization analysis was done by rheology, differential scanning calorimetry (DSC), and principal component analysis (PCA). It was found that changing the stoichiometric ratio of the materials system allows the molecular weight and rates of crystallization to be judiciously controlled. Also, increasing the temperature at which the reaction takes place was appropriate for controlling the crystallization rate. By adjusting these variables, the processes of polymerization and crystallization - which normally are coupled - were temporally separated as necessary for AM. Furthermore, it was found that adding Cr nanoparticles to the material system enabled induction heating with a simple radiofrequency field to be used to melt the linear polymer uniformly and rapidly. Overall, using photopolymerization to form this new class of semicrystalline polymers holds great potential in AM and is further enabled by utilizing off stoichiometric mixtures along with 3D printing and induction heating.

638692 Air-Controlled Electrosprayed 3D Graphene/CNT/Noble Metal Nanoparticle Electrodes for Energy Storage Applications

Abstract:

Air-Controlled Electrosprayed 3D Graphene/CNT/Noble Metal Nanoparticle Electrodes for Energy Storage Applications

Evan K. Lee, Sophie R. Cohen, Melanie L. George, Vesa Ibrahimi, Michael J. Williams,

Caspar C. Yi, Enoch A. Nagelli

Department of Chemistry & Life Science, United States Military Academy, West Point, New York 10996

Abstract

The enhanced electrical, thermal, chemical, and mechanical stability of graphene and carbon nanotubes (CNTs) offer an ideal platform for electrode design for energy storage applications.

Here we utilize spontaneous galvanic displacement driven by reduction potential difference to produce three-dimensional (3D) graphene-CNT-noble metal nanoparticle thin films without the use of any chemical reducing agents. A graphene-CNT slurry with a poly(acrylic acid) (PAA) binder is air-controlled electrosprayed onto copper foil to create thin film electrodes. The resulting copper supported graphene-CNT electrodes are immersed into aqueous noble metal salt solutions ($HAuCl_4$, K_2PtCl_4, and Na_2PdCl_4) to enable electroless noble metal nanoparticle deposition onto the 3D graphene-CNT network through a galvanic displacement reaction with copper. Scanning electron microscopy (SEM) imaging confirms that the carbon nanomaterials are integrated with noble metal nanoparticles forming an overall 3D electrode structure. Raman spectroscopy verifies the characteristic D-band, G-band, and 2D-band peaks from the graphitic structure within the 3D carbon and noble metal nanostructure. Cyclic voltammetry and electrochemical impedance spectroscopy (EIS) are used to characterize the electrochemical properties of the electrodes. We demonstrate that the use of an energy-free and spontaneous process based on the difference in thermodynamic reduction potentials as the driving force for producing carbon nanomaterial/noble metal nanostructured electrodes for batteries and fuel cells. This process is a more simple, scalable, and cost-efficient alternative to current methods for developing lightweight and catalytic electrodes for energy storage applications, such as lithium-ion batteries, lithium-air batteries, and fuel cells.

KEYWORDS: Graphene, Carbon Nanotubes, Noble Metal Nanoparticles, Batteries, Fuel Cells, Thin Film Electrodes, Energy Storage and Conversion

CONTACT: Dr. Enoch Nagelli, Department of Chemistry and Life Science, United States Military Academy, West Point, New York 10996. Email: enoch.nagelli@westpoint.edu

638473 Studying the Fundamental Mechanics behind Cesium Lead Iodide Perovskite Phase Degradation Due to Surface Defects

Abstract:

Currently, the solar cell market is dominated by silicon based solar cells. However, over the last decade, photovoltaic cells made using halide perovskite absorbers have shown rapid and consistent increases in efficiency. This, coupled with their improving stability and inexpensive manufacturing, has brought much attention to the material as a potential competitor to silicon as both a semiconductor more generally and a solar cell technology. There is a desire to improve thermal stability of halide perovskites, which have the general formula ABX_3, by using cesium lead iodide ($CsPbI_3$); however, crystal phase stability for $CsPbI_3$ based perovskites is poor. This phase stability is easily seen visually as the perovskite film will transform from a dark brown/black perovskite phase to a visibly transparent clear non-perovskite phase. One suspected factor influencing the metastability of the perovskite phase in $CsPbI_3$ is the surface defect concentration, specifically iodide vacancies. During high temperature thermal annealing, defects form on the crystal surface where, it is hypothesized, they become nucleation sites for the non-perovskite phase to begin the phase transformation. In this study, $CsPbI_3$ perovskite film solutions were made and post-treated with increasing concentrations of iodide by the addition of CsI and CdI_2 to reduce the concentration of surface iodide vacancies. Two distinct film formation methods were used to fabricate $CsPbI_3$ to understand the generalizability of the results. In the first method, a stoichiometric mixture of CsI and PbI_2 was dissolved in dimethylformamide and deposited by spin-coating. In the second method, a mixed dimethylformamide/dimethyl sulfoxide solvent system was used and an antisolvent, methyl acetate, was used during spin-coating to improve film formation. The phase transformation between the perovskite and non-perovskite phases was monitored at constant temperature and humidity using a home-built system to control film conditions and monitor absorbance. From this, we find that reducing iodide defect concentrations using both CsI and CdI_2 treatments directly leads to improved phase stability, providing strong evidence that iodide vacancies are nucleation sites for non-perovskite phase formation in $CsPbI_3$ perovskites. This insight has important implications for materials design for phase stable halide perovskite materials.

638666 A Systematic Study of the Impact of Charge Capacity and Current Density on the Reported Coulombic Efficiency of Lithium-Copper Batteries

Abstract:

Coulombic Efficiency is a characterization used by researchers to assess cyclability and practicality of new and/or improved batteries. Mathematically, coulombic efficiency is the ratio of the total charge extracted from the battery to the total charge input into the battery over a full cycle. This characterization thus describes the charge efficiency by which electrons are transferred in batteries and can thus be used to effectively indicate battery performance.

However, different researchers use different charge capacities and current densities in their battery cycling systems to assess coulombic efficiency. Consequently, it is important to investigate whether or not the parameters chosen to cycle batteries will influence their reported coulombic efficiency.

This poster will present the findings from a systematic study where lithium-copper coin cells of identical compositions are cycled under varying charge capacities and current densities and their coulombic efficiencies assessed. The coin-cells used for this study are assembled using commercial grade celgard battery separators, with the aim that this study could be used to improve the assessment of practical commercial batteries which incorporate these types of separator. Additionally, both carbonate and ether electrolytes were incorporated in the assembly of the coin-cells used in the study, so that the findings may be applied to a larger range of battery compositions.

The findings of this study indicates that the choice of current density and charging capacity in the cycling of batteries of the same composition does in fact affect reported coulombic efficiency. Consequently, a standard charge capacity and current density should be used by researchers to assess reported coulombic efficiency, so that findings from different studies may be fairly compared without results being skewed based on individual choices of current density and charge capacity.

638462 Synthesis of Post-Spinel Compounds for Calcium Ion Battery Cathodes

Abstract:

Rechargeable batteries, which currently rely on lithium ion technology. are widely employed in electric vehicles and phones. As demand for rechargeable batteries grows, calcium ion batteries represent a potential replacement for Li-ion batteries as concerns grow regarding the price, availability, and environmental impact of Li-ion technology. Ca-ion batteries have potential applications in supporting intermittent renewable energy sources such as wind and solar power. Calcium has advantages over lithium because of its relatively high abundance which mitigates the scarcity and high cost issues surrounding lithium. Additionally, calcium is more sustainable and has less of an environmental impact. In terms of battery performance, calcium and lithium have a similar volumetric capacity and the divalent nature of calcium promises a potentially greater energy density. Commercial scale production of Ca-ion batteries has been prohibited by a lack of suitable cathode materials. Calcium oxides in the post-spinel phase are of interest for cathode materials owing to their probable high ionic mobility. DFT calculations indicate a theoretically low energy barrier for Ca-ion migration in the post-spinel structure. Three post-spinel compounds, $CaFe_2O_4$, $CaMn_2O_4$, and $CaCo_2O_4$, were synthesized using varying methods. X-ray diffraction results indicate a high purity of the post-spinel phase in each synthesized compound compared to commercially available materials. Future work includes the processing of the synthesized post-spinel compounds into electrodes and evaluating their redox activity with electrochemical testing.

638412 Study of the 2,5-Dihydroxy-1,4-Benzoquinone As Redox-Active Material for Aqueous Alkaline Redox Flow Battery

Abstract:

Aqueous organic redox flow batteries (AORFBs) are devices used to store energy. They consist of organic active species dissolved in an aqueous electrolyte (negolyte and posolyte), which are held in separate external tanks. The species react reversibly when they are pumped through the cell electrodes and transfer electrons through an external circuit. AORFBs offer an advantage over an all-vanadium redox flow battery, the most commercialized inorganic-based RFB, because their organic redox material is more earth-abundant and has a potentially lower cost. In this work, we studied the usage of 2,5-dihydroxy-1,4-benzoquinone (DHBQ) as an organic redox-active molecule in alkaline electrolytes. Despite previous work that has shown that DHBQ has a high potential as redox-active material for RFB, some problems associated with membrane cross-over, that generates a loss of capacity, and electrolyte stability need to be addressed. Thereby, we report our results of a study with different ion-exchange membranes and our current studies on the stability of the DHBQ in alkaline media.

638521 Analysis of Electronic Properties of $FAPb_{1-X}Sn_XI_3$ Perovskites

Abstract:

Pb-based hybrid organic-inorganic perovskites, such as $MAPbI_3$ and $FAPbI_3$, are among the most promising materials to be widely used for light absorption in solar cells. However, Pb-based materials pose safety hazards and cannot be tuned for use as rear cell absorbers in tandem solar cells. Completely substituting Pb with Sn allows lowering the band gap, but also undermines the stability of the materials. In this study, the effects of alloying $FAPbI_3$ with Sn in place of Pb at different ratios are examined in terms of the band structure, density of states, and charge density, using density functional theory (DFT) methods. It was found that substituting Pb with Sn does not significantly change the structure or dimensions of $FAPbI_3$, and it generally lowers the band gap. The band gap is lowered due to the contribution of Sn to the conduction band minimum of the material. Due to the band gap tunability through alloying, hybrid organic-inorganic perovskites show promise as rear cell absorbers in tandem solar cells.

638263 Simulation and Analysis of Sulfone-Based Localized High-Concentration Electrolyte Systems for Lithium-Ion Batteries

Abstract:

Lithium (Li)-ion batteries (LIBs) are a state-of-the-art energy storage technology and have been found in many aspects of our daily life, including portable electronics, electric vehicles, and power grid storage. The electrolyte of the LIB serves as the media for the transport of Li ions within the cell and the source for the formation of effective solid electrolyte interphase (SEI). This makes the electrolyte crucial for operation of LIBs. Conventional electrolytes are usually dilute organic-based solutions of Li salt (more specifically, LiPF6), which are often incompatible with high-energy-density LIBs and pose safety risks due to their high-flammability. Li metal batteries (LMB), cells with Li metal anode, the optimal electrode, also deplete conventional electrolytes rapidly. Localized high concentration electrolytes (LHCEs) have recently emerged as a promising class of electrolytes for LIBs and LMBs, utilizing the electrochemical stability and safety benefits associated with high concentration electrolytes, but also sustaining improved viscosity and electrode contact. Sulfone-based LHCEs can utilize the benefits of good stability on high-voltage cathodes, high dielectric constant, and low flammability associated with sulfone solvents. They can also solve the issues of high viscosity and poor wettability associated with sulfone-based high concentration electrolytes. Additives can offer increased performance and improve SEI stability. Greater understanding of microscopic interactions within LHCE systems is critical to explaining many of their macroscopic properties. Here, we describe the use of molecular dynamics to investigate the theoretically predicted properties of tetramethylene sulfone based LHCEs with additives, such as solvation structure and dielectric constant, and compare them with experimental measurement. We also present microscopic images and electrochemical cycling data demonstrating their high performance in LIBs. Our investigations will lead to greater fundamental understanding of LHCEs and offer first principle guidelines for developing advanced electrolytes for rechargeable Li batteries.

638003 Demonstration of a Flexible Pressure Sensor with Novel Piezoelectric Materials

Abstract:

Piezoelectrics are materials that have the ability to generate electrical charges when an outside stress is applied. Furthermore, the research into alternative piezoelctrics can be essential in the production of many commonly used products like hand lighter and pace makers. The piezoelectric properties of a material can be measured by running a high voltage current through the material to align its dipoles and testing the electric response the material can produce when pressure is applied. Some organic and inorganic materials are naturally piezoelectric, and are also inexpensive to produce. This would allow for manufacturers to move away from PZT (Lead Zirconium Titanate), which possesses strong piezoelectric effects, but the lead found within it is extremely hazardous to society and the environment. Here we show that PVDF (Poly(vinyldene)difluoride), a weaker piezoelectric but less hazardous material, plus an organic/inorganic material can be an inexpensive, lead-free, and easy-to-fabricate alternative to PZT. The results confirmed that a combination of PVDF and (Histammonium) $ZnCl_4$ was able to create a functioning active device and produced the best results due stronger piezoelectric effect compared to (Histammonium) $CuCl_4$and (Histammonium) $CoCl_4$. Each material generates an electric signal proportional to the applied pressure. Our results demonstrate how each material can be used to increase the piezoelectric effects of a PVDF polymer. We anticipate that our research will be used as a starting point for other methods of strengthening weaker piezoelectrics as better alternatives.

638334 Mechanistically Studying the Degradation of Cesium Lead Iodide Perovskite Phases Due to Temperature and Humidity

Abstract:

Halide perovskites offer exciting potential as photovoltaic materials and simply as semiconductors. Perovskites' ever-increasing efficiency, stability and lowering manufacturing costs is making them even more attractive as a competitor to silicon solar cells. The focus of this research was on halide perovskites which have the formula ABX_3, and more specifically on cesium lead iodide ($CsPbI_3$). A major drawback of $CsPbI_3$ is its limited stability in the perovskite crystal phase, the colored phase with desirable optoelectronic properties. The source of this instability is a size mismatch between the cesium cation and the lead iodide octahedra. Cesium is somewhat too small of a cation to stabilize the perovskite structure and in the presence of outside factors, such as increased humidity and temperature, the metastable perovskite structure changes to the thermodynamically favored, mostly colorless orthorhombic phase. Most work has centered on improving the stability of the perovskite phase whereas less has been done to mechanistically study the phase change process. In this study, a home-built system was constructed to analyze the absorbance of a $CsPbI_3$ film over time while controlling humidity and temperature. Absorbance data collected was converted to phase fraction data and the phase transformation was modeled by the JMAK model, $x(t) = \exp(-kt^n)$, where n is the growth coefficient, k is the rate constant and t is time. It is found that relative humidity increases the phase transformation rate exponentially, indicating a first order process; however, increasing temperature leads to a monotonic decrease in the rate constant. We hypothesize that the decrease in phase transformation rate with increasing temperature is the result of surface water desorption. Experiments are also consistent with the phase transformation rate being nucleation rate limited where individual grains transform very rapidly once a non-perovskite phase nucleates, but neighboring grains do not influence one another. Work continues to model the adsorption and desorption mechanism of water onto the surface of $CsPbI_3$ and the effect this has on the phase change by analyzing the effects of differing temperature and humidity.

638178 Energy Storage with Seawater Electrolysis: Ability of Oxide Coatings to Manipulate Selectivity Towards Oxygen/Chlorine Evolution at a Buried Interface

Abstract:

Energy is a globally valued resource. In order to meet energy demands in the face of climate change, developing efficient methods of storing renewable energies is essential. A clean method to store energy is through the electrochemical splitting of water. Such a reaction can be driven by renewable energy sources like solar energy. In this project, an abundant natural resource is explored as an electrolyte for electrolysis—seawater. The electrolysis of seawater evolves hydrogen and oxygen gas, which can act as fuels when recombined in a fuel cell to generate electricity, heat, and desalinated water. This technology holds promising potential for the clean production of hydrogen and desalinated water. The major benefit of utilizing seawater as a feedstock is the preservation of fresh water sources. However, the high concentration of chloride salts in seawater poses challenges, resulting in competition between the oxygen evolution reaction (OER) and chlorine evolution reaction (CER) at the anode. Our goal is to understand and control an electrode's activity and selectivity towards OER in seawater electrolysis. To accomplish this goal, electrodes with varying thicknesses of manganese oxide (MnOx) were electrodeposited onto iridium oxide (IrOx) coated glassy carbon substrates. The performance of each electrode in neutral chloride solution was compared through rotating ring-disk electrode cyclic voltammetry. From initial testing, MnOx decreases CER selectivity. While further testing is needed, initial tests indicate increasing thickness of MnOx films increases selectivity towards OER. This understanding informs future design of better electrodes to advance seawater electrolysis technology for hydrogen production.

637627 In-Flow Facile Cation Doping of Metal Halide Perovskite Quantum Dots

Abstract:

Metal cation doping of all-inorganic metal halide perovskite (MHP) quantum dots (QDs) enables incorporation of new optical, optoelectronic, and/or magnetic properties into the pristine QDs. In this work, we present an ultrafast (< 2 s) metal cation doping strategy for all-inorganic MHP QDs. Specifically, we investigate the role of dopant concentration and ligand-to-solvent ratio on the kinetics and extent of metal cation doping of an exemplary MHP QDs, cesium lead chloride, using a modular microfluidic platform equipped with a translational spectral monitoring probe. Manganese (II) chloride, Oleylamine, and 1-octadecene were selected as the source of the metal cation dopant, ligand, and reaction solvent, respectively. We illustrate that the emission color of the doped MHP QDs can be tuned within the blue-violet to orange-red regions through tuning the dopant concentration. The in-flow cation doping of MHP QDs was precisely monitored through real-time in-situ UV-Vis absorption and photoluminescence spectroscopy. The unique translational spectral monitoring probe coupled with the time-to-distance transformation enabled by flow chemistry allowed us to monitor the progress of the metal cation doping at various residence times (*i.e.*, reaction times). The benefits of conducting the metal cation doping of all-inorganic MHP QDs in flow, compared to batch reactors, are enhanced and precisely tunable mass transfer rates as well as time and resource savings. In addition, in-flow studies of colloidal synthesis and doping of QDs allows for faster and more predictable results, allowing for future prospects of on-demand precision synthesis of metal-cation doped QDs with desirable optical and optoelectronic properties for optoelectronic and photonic devices.

638685 Emulsion-Induced Polymersomes for Water-Soluble Drug Encapsulation and Sustained Release

Abstract:

The effectiveness of water-soluble drug delivery is limited by low carrying capacity, undesirable load leakage. The research aim is to develop emulsion-induced polymersomes capable of encapsulation and sustained release of water-soluble drugs. We systematically altered the composition of each polymer and tested each sample for carrying capacity and stability, and controlled release trend. The hypothesis was that a more soluble molecule would increase carrying capacity due to the ease of forming of aqueous pockets, and have higher stability due to a smaller repellent force between the hydrophilic drug and the polymer shell. The oil based polymers used were HPG-PPG-HPG and PPG-HPG-PPH at different levels of H branching, and thus solubility.

The tests performed on the emulsion induced polymersomes of varying branching were a 24 hour stability test in vitro, and a 36 hour release test for release in vitro. We also examined these particles through a microscope and NMR, comparing the results to predictions made through software to confirm the structure was as intended.

We found that hyperbranched polymers did have higher load capacity and stability as compared to molecules with smaller branching, most likely due to the high hydrophilicity, which prevented collapse and allowed larger pockets of the aqueous payload to form more easily inside the shell.

638257 Novel High-Modulus Cell-Delivery Composite Biomaterial for Intervertebral Disc Repair and Regeneration

Abstract:

Lower back pain is the most common cause of disability, which domestically costs $134.5 billion per year. Back pain is strongly associated with intervertebral disc (IVD) pathologies like herniation, but the current standard of care for IVD herniation, discectomy, does not repair defects in the outer annulus fibrosus (AF) structure caused by the initial herniation. Defects in the AF can lead to accelerated degeneration, recurrent pain, and reherniation, highlighting a need to develop therapies capable of promoting IVD repair. Injectable cell delivery biomaterials are an attractive solution because they may provide immediate biomechanical stabilization after injury and deliver cells for long-term healing. However, research is required to develop high-modulus biomaterials that can be retained within the high loading environment of the IVD while supporting the function of delivered cells.

We developed a composite biomaterial for AF repair that uses cell-seeded, degradable oxidized alginate (OxAlg) microbeads to deliver AF cells seeded within high-modulus genipin-crosslinked fibrin (FibGen) hydrogels. The goal of this study was to evaluate the ability of this composite cell delivery biomaterial to repair bovine coccygeal IVDs in an *ex-vivo* loaded organ culture bioreactor system. We hypothesize that this composite will resist herniation, prevent disc height loss, and promote repair.

IVD viability in the organ culture system was confirmed by constant, low levels of nitric oxide and lactate dehydrogenase in the culture media at days 7, 21, and 42. Herniation risk was assessed by visual inspection of the repaired IVDs. The composite biomaterial remained in place after 96,000 cycles of compressive loading over 6 weeks without evidence for herniation. This herniation test is rigorous since very few *ex-vivo* studies test experimental IVD repair biomaterials for this many cycles or as extended a duration. Analysis of disc height change, the most sensitive metric for assessing IVD injury and repair, showed that injured IVDs had significantly greater disc height loss after 42-day culture than intact controls. Excitingly, repaired IVDs experienced significantly less disc height loss than injured IVDs, which was similar to intact controls. Future analyses will assess stiffness, track the fate of delivered AF cells, and measure extracellular matrix synthesis.

Overall, this work shows that FibGen hydrogels with AF cell-seeded OxAlg microbeads restore lost disc height without herniation and may offer promise for AF repair. This cell delivery composite, designed to balance biological and biomechanical performance, may also have relevance to repairing other tissues with limited regeneration and high mechanical demands.

638330 Tailored Drug Release from Polymeric Coatings Fabricated Via Electrospray

Abstract:

Electrospraying is an effective technique for producing polymeric coatings on conductive substrates with smooth topographies. This technique is unique based on its ability to create reproducible coatings in ambient conditions from a plume of charged droplets. These polymeric coatings can be homogeneously fabricated to incorporate potential drug molecules for controlled drug delivery. Conductive substrates such as stents can be coated and used as a vessel for localized drug release in the body. The main objective of the study is to create drug loaded films that exhibit controlled release into a media, Phosphate Buffered Saline (PBS). *In vivo* conditions were simulated by utilizing a thermal shaker and replenishing the PBS after each model drug concentration measurement. The total cumulative percentage of drug release was determined for different model drugs. The controlled release observed was a result of exploiting the change in the glass transition temperature of the polymer, Polyvinyl Acetate (PVAc). PVAc coatings were deposited on metal disks with known geometry to act as vessels. The release kinetics of various coating thicknesses between 10-40 micron have been analyzed. Tailored drug release is desirable for many medical treatments, including cancer treatment.

638200 Effect of Weight Ratio and Casting Method on the Release of Drugs from PEO-*B*-PCL Copolymer Films

Abstract:

Current methods for drug release use PCL encapsulation. PCL is used because it is biocompatible and has low toxicity within the body; however, the polymer is very hydrophobic, taking up to 2 years to degrade. The aim of this study was to create polymer films with a PEO-*b*-PCL copolymer and determine how the hydrophilic PEO affects the amount of drug released as well as the time it takes for the film to fully erode. This was tested using three different dyes as drug surrogates: methyl red (MR), methyl orange (MO), and methylene blue (MB). The effect of the film casting method was investigated by making the films using two different methods: a compounding method and a solvent casting method. The films were placed in water and samples were taken at regular time intervals. The dye release was determined by UV-vis spectrophotometry. Initial studies showed that the more PCL present in the film, the slower the release of the dye. In general, initial studies show that samples using MB degraded at a faster rate while samples using MR degraded at a slower rate. Also, it was found that the solvent method results in a faster release.

638013 Electrochemical Diazonium Grafted Surface for Protein Biosensing

Abstract:

Surface-based or electrochemical protein biosensors commonly use thiol-gold bonding to conjugate a designed binding protein (ligand) to the surface in order to evaluate analytes in the solution. As an example, antibodies, proteins, and other large polypeptides have been conjugated to gold surfaces that lead to successful biosensors. While thiol-gold surfaces have been shown to lead to successful biosensor surfaces, they have limited stability and long-term use, which can affect overall sensor performance. For example, thiol-gold surfaces can degrade (thiol decoupling) or the gold can become pitted. Surface damage in such forms presents numerous problems when paired with electrochemical impedance spectroscopy (EIS) or cyclic voltammetry (CV) when concerned with long term sensor stability.

To overcome the degradation of thiol-gold bonds, we created a more stable surface by electrochemically grafting diazonium functionalized surfaces. To build the surface, 4-carboxyphenyl diazonium salt was electrochemically grafted to a carbon surface creating a functional carboxylate. The N-terminus of a polypeptide is modified to the carboxyl-terminated surface via EDC/NHS chemistry. The surface was characterized using XPS, ATR-FTIR, and EIS. Long term surface stability of the 4-carboxyphenyl surface compared to thiol-gold surfaces indicated a stability increase over a one month period. The stability and viability of diazonium/protein surfaces has also been investigated and will be reported.

638183 Menstrual Health and Hygiene

Abstract:

The *Menstrual Health and Hygiene (MHH)* project focuses on the development of a biodegradable and compostable menstrual pad for underserved communities. The project is a collaboration with the OSU Global Health program which has an ongoing relationship with the village of Muanatlala, Botswana (approximately 5000 people) that is the initial target community. The MHH Team been testing many types of materials and designs to find the best construction of an *effective* and *compostable* menstrual pad. An additional consideration is that there is an approximately 30% HIV rate amongst the village girls and women. Menstrual pads in industry are judged on their strikethrough, absorbency, wetback and adhesion. Using these standards with the addition of the need to be compostable in local environments such as Botswana, materials have been sourced that meet the criteria. Ongoing research involves testing of both the materials and construction design of prototype menstrual pads. Once the final prototype design is selected, the next phase is manufacturing in-house 1500 menstrual pads for delivery to Botswana in summer 2022.

637928 N-Acetylcysteine and Hydrogen Sulfide Releasing Biomaterials for Neuronal Protection and Differentiation in Peripheral Nerve Injury Regeneration

Abstract:

In the United States, twenty million Americans suffer from peripheral nerve injury (PNI) due to trauma and medical disorders. The majority of these injuries are a result of upper limb trauma and can result in considerable physical disability and long term decrease in quality of life. This is partly due to the several limitation surgical treatments (e.g. direct nerve grafts, nerve transfer, etc) for neural injuries that are often associated with sub-optimal functional recovery with fewer than half of the patients who receive nerve repair procedures post-injury regain complete mobility and sensitivity. This highlights the inadequacy of present-day treatment methods. The use of hollow nerve guidance conduits for nerve repair is clinically approved as an alternative treatment to the "gold standard" autologous neural grafts for PNIs.

Recent efforts in the field have shifted towards exploring alternative solutions such as neural scaffolds that incorporate biological cues. Simple signaling molecule Hydrogen Sulfide (H_2S) has been shown to possess neuroprotectant and neural differentiation capability that could assist in the healing of PNI *via* hollow nerve guidance conduits. Another promising neuroprotectant is N-acetylcysteine (NAC), a known antioxidant derived from L-cysteine and is medication used to treat paracetamol (acetaminophen) overdose, and that has demonstrated capabilities in suppressing mitochondrial reactive oxidative species (ROS) by increasing intracellular Glutathione (GSH) concentration.

This research looks therapeutic capability for H_2S and NAC through cell proliferation and preliminary efforts to synthesize monomers thioglutamylglutamic acid (GluGluSH) and N-acetylcysteineglutamic acid (GluNAC). This was accomplished through cellular studies using neuroblastoma (NE-4C) cells exposed to either NAC or NaSH and H_2O_2. DNA, ATP assays and fluorescent microscopy were used to determine therapeutic window and capability of both NAC and NasH. Synthetic work done includes thiolation reaction of Boc-Glutamic (OBzl) acid and sequential palladium on carbon reduction reaction. Products formed were detected using thin-layer chromatography (TLC), purified using column chromatography, and characterized for their structure confirmation by employing Fourier transform infrared spectroscopy (FTIR) and proton (^1H) as well as carbon (^{13}C) nuclear magnetic resonance (NMR) spectroscopy techniques.

Future research will focus on continued synthesis of GluNAC and GluGluSH monomer to then leverage for various degradable polymers (e.g, polyanhydrides and polyester), for which cell studies can be guided by the already established therapeutic window. Ultimately, these neurogenerative and neuroprotective biomaterials have the potential to be integrated in hollow nerve guidance conduits for treatment of peripheral nerve injuries.

638108 Assessing Self-Nucleation to Promote Minority Block Crystallization in Poly(ethylene oxide)-b-Poly(ε-caprolactone) to Control Degradation Rates for Drug Delivery

Abstract:

PCL is used in drug delivery, however, the degradation rate is too slow for it to be truly effective. When PEO is added to PCL, it enhances the degradation process. PEO and PCL are both crystalline materials, and the crystallinity affects the degradation properties. This study evaluated self-nucleation temperature changes for PEO-*b*-PCL This was accomplished by using a Differential Scanning Calorimeter (DSC) to observe the crystallization temperatures of samples with different weight fractions and solvent casting techniques. It was found that both have an effect on the self-nucleation properties of PEO-*b*-PCL. Going forward, identification of crystallization order after self-nucleation is critical to understanding the crystallinity of these samples.

638052 Characterizing Stability Conditions for Coiled-Coil Formation

Abstract:

In Nature, complex biopolymer molecules help form the basis of life. Proteins and polypeptides are made up of sequences of amino acids, which define the structure and function of the molecule. The primary sequence specificity and its associated assembly into higher ordered structures provides a framework for engineering of materials. One higher ordered structure is the coiled-coil motif, which has a variety of desired properties as a molecular building block. The coiled-coil can be designed to be tolerant to a variety of solution conditions and can be modified with a variety of chemical handles. Characterization of the thermodynamic properties that underlie the coiled-coil stability is critical for understanding when and how these motifs form and provide insight into more reliable peptide design. Here, the temperature dependent assembly/disassembly of peptides to form coiled coils are assessed over a range of solution conditions using circular dichroism (CD) spectroscopy. The effect of both peptide concentration in water and salt types of different concentrations on the melting temperature was assessed. For the computationally designed coiled-coil sequence that was selected, the melting temperature was found to be independent of coiled-coil concentration from 25 μM to 200 μM, while up to 70 mM concentrations of salts enhanced the melting temperatures. Future work includes the evaluation of salts over different valences as well as pH conditions.

637847 Creation of Shear Flow Wash System for Cellular Micropatterning with Biotin-Streptavidin

Abstract:

In the medical industry, organs and tissue are continuously in demand to maintain the health of patients. There has been a rise in the research and production of artificial tissues and organs to generate resources to meet the necessary demands. The artificial tissues maintain regenerative properties through the use of cells placed on the construct. Surface engineering is used to intentionally place cells in a biomimetic pattern on artificial tissue constructs. Cells were placed onto a substrate by chemically altering the surface with strong binding forces like biotin and streptavidin. These molecules create a specific region for cells with biotin molecules on the surface to bind and generate a desired pattern. Through the use of a parallel plate flow chamber, the substrate with the generated cellular micropattern can be rinsed to remove any cells remaining on the substrate that is not bound to the patterned regions. Using a flow chamber and 25 ml/min flow rate with an exposure time of 15 seconds a clear cellular micropattern can be generated.

638690 The Role of Surface Hydroxyl Groups in the Thermal Transport of Silica Aerogels

Abstract:

Concentrated solar thermal (CST) is a promising carbon-free technology which generates high-temperature heat by using large optics to focus sunlight onto a solar receiver. One major engineering challenge is to develop selective receivers that capture sunlight but limit thermal losses due to thermal re-radiation. Silica aerogels are a great candidate for providing insulation to reduce these losses due to its low density and high solar transparency. However, the performance of silica aerogels is susceptible to changes in relative humidity. Current literature suggests there is a linearly proportional relationship between increasing heat capacity of silica and thermal transport. Here, it is demonstrated there is also a linear relationship between the surface concentration of hydroxyl groups and heat transfer properties. We subject the aerogel to relative humidity greater than 70% while measuring the optical properties and surface chemistry of silica aerogels. To do this, we use UV-Vis and FTIR spectroscopy, and transient plane source conductivity analysis. Our results indicate there is a direct relationship between water concentration and thermal transport, giving insight on the nanostructure of silica aerogels and its potential for long term use in CST systems.

638600 Electrochemical Characterization and EPR Studies of TEMPO-Based Electrolytes for Redox Flow Batteries

Abstract:

Redox flow batteries are energy storage devices capable of holding large amounts of energy in comparison with traditional batteries. These batteries utilize redox species that are dissolved in an electrolyte and stored in two external tanks, making it possible to decouple power and energy, making them ideal for storing electricity produced solar and eolic technologies, where the energy production depends greatly on the fluctuations of the source. State-of-the-art redox flow batteries use vanadium solutions as electrolytes. However, vanadium is expensive and has to be dissolved in concentrated acid solutions to obtain the best performance, which poses safety concerns. In this study, we investigated the physical properties and electrochemical behavior of the newly synthesized (2,2,6,6-tetramethylpiperidin-1-yl)oxyl (TEMPO) derivatives. Electron paramagnetic resonance (EPR) gave us insight into the molecular dynamics of the TEMPO derivatives, while cyclic voltammetry and the rotating disk electrode techniques helped us to determine the electron transfer kinetic process of the molecules and the stability of reaction products. The results show that the studied TEMPO-based electrolytes are promising candidates for redox flow battery applications.

638699 Perspectives on Chemical Stability and Corrosion of Magnetocaloric Materials

Abstract:

Active magnetic regenerative (AMR) refrigeration is an energy-efficient and environmentally-friendly alternative to conventional vapor-compression refrigeration technology associated with harmful chemical refrigerants and high carbon emissions having high ozone-depleting potential that encourages planetary warming. To address this challenge and in response to strategic global treaties such as the Montreal and Kyoto Protocol, many countries worldwide, including the European Union (EU), Japan, USA, and China, have begun to unveil new rules to phase out GWP gases. These policies have inspired collaborative research efforts worldwide to find possible alternative cooling methods. Magnetic cooling devices enabled with the "*magnetocaloric*" class of functional materials are attractive as they have the potential for efficiency improvements of up to 50% over conventional vapor compression systems, which is equivalent to 60% of Carnot efficiency. The core component of AMR is a porous magnetocaloric material (MCM) that undergoes millions of thermal and magnetic field cycles throughout the device's lifetime while immersed in a heat transfer fluid. Despite significant research on MCMs spanning three decades, an overlooked and less frequently studied but critical engineering challenge to be considered is the interaction between the magnetic refrigerant material and the heat exchange fluid, namely, the aging of the material when exposed to prolonged corrosive action. Research on the corrosion and chemical stability of room-temperature MCMs is discussed with particular attention given to Gd, $Gd_5Si_2Ge_2$, $La(Fe,Si)_{13}$ and its compositional variants. Following a brief overview of the wide variety of corrosion monitoring methods used to evaluate magnetocaloric regenerator structures, corrosion inhibition mechanisms are discussed in the context of metallurgical, processing, and environmental factors. Challenges associated with corrosion testing of magnetocaloric alloys prepared by additive manufacturing, corrosive-dependent magnetic data on 3D printed $La(Fe,Si)_{13}$ magnetocaloric structures, and insights into corrosion resistance of 3D printed magnetocaloric structures are presented.

638564 Understanding the Oxidation Behavior of Polymer-Derived Si(O)C and Doped $NbSi_2$ Ceramics

Abstract:

Recently there has been an increased demand for high temperature structural materials whose properties exceed those of conventional Ni-based alloys and monolithic ceramics. Therefore, this project, examines doped $NbSi_2$ and SiC-based polymer-derived ceramics that could show improved resistance to environmental degradation at elevated temperatures. Al, Cr, and Ti-doped $NbSi_2$ ceramics, are be made by spark plasma sintering (SPS) elemental powders followed by annealing. For the SiC specimens, SiC powder was mixed with 1 vol% $TiSi_2$ powder, and pressed into pellets. The TiSi2 functions as an embedded marker to track oxygen activity. The pellets were infiltrated with an allyl-hydrido-polycarbosilane (SMP-10) polymer precursor. Subsequent pyrolysis in argon gas at 900°C transforms the infiltrated SMP-10 into amorphous Si-(O)-C. The oxidation resistance of the ceramics is explored in temperature ranges of 1100-1400°C inside a tube furnace under a constant flow of dry air. Cross-sectional Scanning Electron Microscopy (SEM) and Energy-Dispersive Spectroscopy (EDS) analysis are used to characterize the oxidation resistance in both material systems. For the $NbSi_2$ samples, the mass change per surface area was measured, the results yield a value near to the goal of mass gain less than 2 mg/cm^2. In the Si(O)C specimens, crack formation during processing was first significantly minimized. Oxidation treatments illustrated that $TiSi_2$ can be used to gauge the partial pressure of oxygen, which can help understand oxygen transport in Si(O)C.

638382 Exploring Halide Perovskite Structural Tunability to Design Materials for Dynamic Photovoltaic Windows

Abstract:

Halide perovskites offer exciting potential as photovoltaic materials and simply as semiconductors. Specifically, their structural tunability has become of greater interest as researchers begin to search for novel ways to tune the materials to achieve improved solar cell stability or to target new applications. One potential technology which halide perovskites could enable is dynamically switchable photovoltaic windows: windows which can transition between photovoltaically active (dark) and non-photovoltaic (transparent). We build toward this goal in this work by investigating the intercalation and deintercalation of methylamine gas into 2-dimensional Ruddlesden-Popper (R-P) phase halide perovskites of the type A_2PbI_4. As has been shown with 3D methylammonium lead iodide films, the intercalation of methylamine into the halide perovskite lattice results in a color change to a clear crystalline phase. We find that in some 2-D perovskite systems, deintercalation of the methylamine gas is incomplete, resulting in the formation of secondary phases including n=2 R-P and 3D perovskites, as well as lower dimensional materials; however, other 2-D perovskite phases show reversible intercalation/deintercalation with methylamine, indicating stronger binding between the long-chain ligand and the lead halide octahedra of the 2-D perovskite sheet. This work reveals the relative affinity of various $R\text{-}NH_3^+$ molecules, specifically $R\text{-}C_8H_9\text{-}NH_3^+$ materials such as 4-hydroxy phenethyl ammonium, for the halide perovskite lattice and indicates that templating the 3-D $CH_3NH_3PbI_3$ structure with carefully selected long-chain cations could lead to better reversibility in dynamic photovoltaic windows. Work continues to develop improved guidelines for the design of 2D/3D halide perovskite materials for an array of applications.

638656 Temperature Effects on Salt Dependent Phase Separation in Complex Coacervate Systems

Abstract:

Complex coacervate systems utilize various forces to induce charge driven phase separation under unique conditions. Using the system of strong polyelectrolytes, poly(styrene sulfonate sodium salt) (PSS) and poly(diallyldimethyl ammonium chloride) (PDADMAC), we investigated how temperature impacted phase separation alongside more well known salt curves. Temperature based complex coacervation studies are hoped to lead to future discoveries supporting easier vaccine storage, which became extremely relevant recently with the COVID-19 mRNA vaccines expiring after short times at room temperature. These PSS/PDADMAC coacervates follow lower critical solution temperature (LCST) behavior, separating into two phases with increasing temperature; a dense polymer-rich coacervate phase and a polymer-poor supernatant phase. Water interactions influence the LCST behavior. The goal of this study is to take advantage of the salt-water interactions, and observe the changes it may have on the LCST behavior. We created ion-free polyelectrolyte complexes (PECs) to start with a clean slate in which new salts can be introduced. Salts introduced into the system were NaCl, KBr, NaBr, and KCl. UV-Vis spectroscopy is used to determine the PSS concentration in previously separated and centrifuged samples. These samples are forced back into solution phase by oversaturating the coacervate phase with salt, allowing for the back calculation of the original concentration of the temperature changed sample solution to be observed. These coacervate systems are often studied in the frame of holding and releasing materials through phase separation and solution. We hypothesize that by understanding the role water has on coacervation, we can better understand the driving forces that govern complex coacervation and the effect the salt identity has. This understanding can further the field of materials and polymers for use in temperature sensitive systems.

637488 Effects of Polyisobutylene Additives on the Thermodynamic Properties of a Poly(alpha olefin) Base Oil

Abstract:

Automotive transmission fluids are often composed of a synthetic poly(alpha olefin) base oil modified with additives. The additives serve different functions and improve the performance of the base oils. Dispersants are a type of additive that suspend harmful particles within the oil, preventing their agglomeration. It is however also important to know if addition of the dispersant agent alters other properties of the base oils that may influence their performance as lubricants. Among common dispersants are polyisobutylene (PIB)-based polymers.

In this study, we have explored the effect of PIB (MW = 1986 g/mol) on the density and the related derived thermodynamic properties of a poly(alpha olefin) base oil, PAO 4 (MW = 489 g/mol). We have studied the effect of PIB concentration (1, 2, 5 wt %) at 298, 323, 348, 373, and 398 K over a pressure range from 10 to 40 MPa.

The density determinations were made using a variable-volume view-cell. The cell houses a moveable piston. The position of the piston is recorded with a linear variable differential transformer which allows assessment of the internal volume of the cell under all pressure and temperature conditions. A pressure generator alters the pressure on the back side of the piston, causing the piston to move and therefore changing the internal volume of the cell. A known mass loading allows the calculation of densities at different operating conditions. The view-cell also includes a platinum RTD 150 temperature sensor that is in direct contact with the fluid and a Dynisco pressure transducer to accurately measure the temperature and pressure of the system, respectively.

The measured density values were then correlated with the Sanchez-Lacombe Equation of State (EOS) and the derived thermodynamic properties, namely, isothermal compressibility, isobaric expansivity, and internal pressure, were then evaluated using these correlations [1].

The results show that the base oil modified with 5 wt % PIB displays a distinct decrease in the density while 1 and 2 wt % addition of PIB did not have a significant impact on density. Isothermal compressibility was observed to increase, with a higher increase being displayed for the 1 wt % addition case. Isobaric expansivities were observed to increase, but only slightly, at all addition levels. Internal pressure was lowered at 1 wt % addition case but was observed to increase over that of the base oil for the 2 and 5 wt % PIB.

637804 Effects of Rapid Thermal Annealing on Indium Antimonide Bismide

Abstract:

Mid-wave and long-wave infrared (MWIR and LWIR respectively) photodetection is of vital importance to a variety of applications including astronomical, medical, and military imaging [1]. Current photodetectors operating in this region utilize HgCdTe, a material which contains toxic mercury and often experiences low uniformities that can impede device performance [2]. InSbBi has emerged as a promising alternative to HgCdTe for MWIR and LWIR photodetectors due to the significant redshift in absorption which results from incorporating small amounts of Bi into InSb [1]. Still, InSbBi remains a relatively underexplored material.

This study looks at the effect of rapid thermal annealing (RTA) on InSbBi grown at low substrate temperatures via molecular beam epitaxy (MBE). Annealing is a common post-growth method used on highly mismatched alloys (HMMAs) to improve material quality, specifically for those grown at low substrate temperatures [3]. Early predictions expect InSbBi to be an HMMA. It is also involved in device formation. Many other dilute bismides have shown increases in photoluminescent (PL) intensity post-annealing [3][4][5]. On the other hand, the effect of annealing on the absorption spectra of dilute bismides has been variable. [3][4][5]. In this study, PL measurements were utilized to analyze changes in PL intensity and shifts in the absorption spectrum. Furthermore, as MBE-grown InSbBi is produced at low substrate temperatures, Bi droplets which can impede device performance tend to form. In this study, atomic force microscopy (AFM) is utilized to analyze the effect of annealing on the morphology of Bi droplets.

Annealing InSbBi between 100°C to 250°C was found to significantly decrease PL intensity, and annealing between 375°C to 475°C caused PL to no longer be observable. Furthermore, there were no significant shifts in absorption spectra. AFM measurements were conflicting, though it is possible that Bi droplet size slightly decreased with annealing. Still, droplets were clearly present following ever annealing temperature.

As annealing is a necessary step in device formation, it is valuable to know this step will result in a degradation in optical quality. Future processes for creating InSbBi devices should utilize annealing temperatures of 100°C or lower to prevent PL degradation. Furthermore, alternative methods for improving PL intensity which could compensate can be devised. Additionally, as annealing does not have a significant effect on Bi droplet size, alternatives such as wet and dry etching, should be explored to increase the potential of functional InSbBi-based devices.

637731 Atomic Layer Deposited Boron Nitride Films Act As High Temperature Hydrogen Barriers

Abstract:

Nuclear thermal propulsion (NTP), which can produce twice the specific impulse as chemical propulsion, is limited by hydrogen diffusion. Elemental hydrogen can diffuse into the nuclear fuel elements and cause hydrogen embrittlement through the formation of hydrides, which exert pressure on the surface and cause cracking. Atomic layer deposition (ALD) was used to coat surfaces with thin films of boron nitride of thicknesses ranging from 6-15 nm. Differential thermal analysis was used to prove stability and resistance to hydrogen embrittlement of these coated surfaces in hot (up to 1713 K) hydrogen environments. A high activation energy of hydrogen diffusion through hexagonal boron nitride was calculated to be 3.25 eV. Compared to activation energies of hydrogen diffusion calculated for other thin film materials, it was proven that boron nitride was the most effective material to protect a surface against hydrogen embrittlement.

638147 Use of CO_2 to Trigger Reversible Volume Phase Transitions in Hydrogels

Abstract:

Smart materials are often inspired by natural processes that are responsive to environmental stimuli. CO_2 has garnered interest as a stimulus as it is abundant, inexpensive, and non-toxic. Swelling, or the process of absorbing water, is a fundamental property of hydrogels yet is often not controlled using an environmental stimulus. In this work, CO_2 is used to trigger pronounced, reversible swelling of a crosslinked hydrogel. Specifically, CO_2 responsivity was conferred to hydrogels with a volume phase transition temperature (VPTT), which demarcates the transition from a collapsed to swollen state. We show that these gels display a CO_2-switchable volume phase transition that enables gas-triggered swelling at a constant temperature. To tune the VPTT, compositional studies were conducted to evaluate the effect of varying co-monomer content on the resulting VPTT shifts in these samples. Increasing the concentration of CO_2-responsive moiety resulted in an increase in the initial VPTT as well as a greater shift in the VPTT after CO_2 exposure. Isothermal swelling studies conducted at a temperature between the VPTTs before and after CO_2 exposure show that CO_2 triggers a reversible volume transition and controls swelling. The results presented here demonstrate the promise of CO_2 as a stimulus for the design and study of smart materials, which may find applications in drug delivery, microfluidics, and soft actuators.

638429 Thermochemical Modeling on Graphene Nanosheet Film Exfoliation

Abstract:

Graphene based nanosheet films have gained popularity for use in various applications such as energy storage, selective membranes, semiconductors, and catalysis. Thermal exfoliation is an efficient method for large scale production of reduced graphene nanosheet or reduced graphene oxide (rGO) films. The full thermal exfoliation will lead to the destruction of the original film cast body. As such, partial thermal exfoliation may be used to manufacture rGO films in the original geometry with needed properties. This research provides a simple novel methodology in synthesizing uniform, flexible, and planar rGO films and provides findings acquired through theoretical modeling. The modeling is performed by applying an energy balance on the film during exfoliation. This model considers the heat accumulation, chemical reaction, and the convective heat transfer with the environment. The model is applied to obtain thermal properties and reaction rate law for rGO films. In addition, these constants are used to provide a theoretical sample temperature curve, which has minimal square sum difference with the experimental exfoliation curve. This yields a model that can effectively reproduce various thermochemical processes involving GO thermal exfoliation. Namely, reaction order, activation energy, enthalpy of reaction, specific heat capacity and activation energy were found to be, respectively: 1.59, 196 kJ/mol, 1217 J/g, and 255 J/g K. Modeling GO film thermal exfoliation gives a better understanding of key properties that will aid in fabrication of these rGO films for use in various applications.

638420 Heavy Metals Extraction By Intensifying Method of Emulsion Liquid Membrane with Nanoparticles in W1 Phase & Ionic Liquid in O Phase

Abstract:

A new technique of the Emulsion Liquid Membrane (ELM) enhanced with the combination of nanoparticles (MgO & Al_2O_3) in the internal receiving phase (W1) and ([OMIM]PF6) ionic liquid in the organic phase (O) for removing Pb^{+2} & V^{+5} from synthetic wastewater has been implemented. As a result, Pb^{+2} removing efficacy has been improved noticeably to 95.8% in 3 min for 0.01% (W/W) of MgO NPs in the W1 phase and 5% (V/V) ([OMIM]PF6) ionic liquid in the O phase, and 94.6 % in 3 min for 0.01% (W/W) of Al_2O_3 NPs in W1 phase and ([OMIM]PF6) ionic liquid in the O phase. However, the V^{+5} removing efficacy has been improved noticeably to 98.6 % in 3 min by using 0.01% (W/W) of MgO NPs and 5% (V/V) ([OMIM]PF6) ionic liquid in the O phase, and 97.7 % in 3 min by using 0.01% (W/W) of Al_2O_3 NPs and 5% (V/V) ([OMIM]PF6) ionic liquid in the O phase. In addition, the emulsion stability was enhanced significantly, which gave the lowest leakage to 16% for three days. The conclusion of this work includes the removal of heavy metal & hydrocarbons and polar pollutants from industrial wastewater and water effectively using the newly developed technology of ELM enhanced by nanoparticles and ionic liquid.

638486 Development of Polyelectrolyte Multilayer Membranes for Ammonium Recovery from Anaerobic Digestate

Abstract:

This research aims to develop a membrane separation process to treat anerobic digestate for the recovery of liquid ammonia, which can be used as fertilizer. Industrial manufacturing of ammonia is an energy-intensive process and accounts for a significant release of greenhouse gases. Anaerobic digestate (AD), which contain organic carbons (COD), ammonium, and phosphate, represents a promising resource of nutrients which can be recovered for use as fertilizers.

Membrane separation process can be an effective way to treat such a wastewater feed, provided membranes with tunable organics/nitrogen selectivities can be designed. Our work focusses on developing such a tunable platform using polyelectrolyte multilayer membranes (PEM) by the surface modification of a loose nanofiltration (NF) membrane (NF 270) using layer-by-layer (LbL) deposition methods. The LbL method involves the sequential deposition of alternately charged polyelectrolytes on a given subtsrate for a desired number of cycles. The central hypothesis of this work hinges on the fact the LbL strategy would provide a precise way to control both the water flux across the membranes as well as the selectivity and therefore show better permeability/selectivity trade-offs compared to commercial membranes. Two polyelectrolytes were used for the surface modification: poly (diallyl dimethyl ammonium chloride) (PDAC) and poly (styrene sulfonate) (PSS). The performance of those PEM membranes were compared against a commercially available NF-90 membrane,which is a “tight” NF membrane. Our initial experiments suggest that in comparison with the commercial nanofiltration/reverse osmosis membranes, these PEM membranes exhibited higher rejection of organic carbon with an equivalent phosphate removal. Furthermore, PEM membranes' low ammonia rejection and high water flux with respect to commercial RO membranes make them more feasible to use against AD for ammonium recovery. Beyond the current focus, we aim to establish these PEM membranes as an effective platform for useful resource recovery from varied wastewater sources.

638497 Rational Design of Highly Selective and Plasticization Resistant Polymers of Intrinsic Microporosity (PIMs) Inspired By Competitive Sorption

Abstract:

Diffusion-selective polymers of intrinsic microporosity (PIMs) have shown excellent pure-gas separation performance due to their rigid backbones, inefficient packing, and high free volume. Their out-of-equilibrium packing morphologies, however, make PIMs susceptible to physical aging and their intrachain rigidity alone has shown insufficient to mitigate plasticization, where selectivity is significantly reduced for gas mixtures at high-pressure. A number of studies on the mixed-gas transport of PIMs have shown the effects of competitive sorption on separation performance, where gases with high polymer affinity (e.g., CO_2) can reduce the sorption of co-penetrants in a mixture (e.g., CH_4, N_2) and increase sorption selectivity. However, due to CO_2-induced plasticization at high pressures, decreases in diffusion selectivity can outweigh beneficial competition effects. This trade-off in performance is especially detrimental for PIMs with little CO_2 affinity and poor plasticization resistance, as they rely primarily on diffusion-selective transport that may be significantly reduced at high pressure.

Herein, we report on mixed-gas and high-pressure transport properties for six PIMs with identical benzodioxane backbones and a diverse set of backbone functionalities. Low-pressure mixed-gas tests indicate a relationship between CO_2 sorption affinity and enhancements in CO_2/CH_4 mixed-gas selectivity compared to pure-gas for all PIMs considered. The best results are reported for amine-functionalized PIM-1 (PIM-NH_2), which shows an unprecedented 140% and 250% increase in CO_2/CH_4 and CO_2/N_2 mixed-gas selectivity, respectively, compared to that of pure-gas tests at 2 atm. Moreover, PIM-NH_2 films retain high mixed-gas selectivity (>20) up to a total mixed-gas pressure of 26 atm in 50/50 CO_2/CH_4 mixtures, demonstrating strong plasticization resistance. Pure-gas sorption and mixed-gas permeation performance for the six PIMs were compared across a range of reported microporous polymers, elucidating structure/property relationships that can enable rational design of polymers capable to perform well in industrially relevant scenarios. Results demonstrate the promise of primary amine functionalization for developing highly sorption-selective and plasticization-resistant membranes for gas separations.

638644 Engineering PFAS Membrane Adsorbers

Abstract:

The contamination and availability of safe, potable water is a concern affecting individuals globally. Poly- perfluoroalkyl substances (PFAS) are an example of organic contaminants found in public drinking water systems, natural waters systems, wastewaters, and sludge. This contamination dilemma is attributed to the use of PFAS in a variety of applications industrially and commercially, as the waste can contaminate surface and groundwater. PFAS are composed of a hydrophobic fluorinated carbon chain tail and a hydrophilic head group that is typically a carboxylic or sulfonic acid, and thus offer desired surface properties, such as a water and oil repellent. PFAS are utilized in products ranging from surface coatings, fabrics, non-stick cookware, and fire-fighting foams. Although they allow excellent surface properties, PFAS accumulate in water systems over time due to their inability to degrade by natural elements and most, if not all are toxic. Additionally, PFAS are resistive to heat, harsh chemical conditions, and water, therefore, make them difficult to remove with water and wastewater treatment plants. Current methods for the removal of PFAS from contaminated water include absorption using granular activated carbon (GAC), ion exchange resins, and membrane filtration using reverse osmosis (RO) and nanofiltration (NF). However, due to the high energy required and concentrated waste streams produced, as well as the long process time of GAC and ion exchange resins, the use of a porous polymer membrane is a more efficient method of PFAS adsorption. Careful selection of absorber chemistry allows for potential regeneration, unlike GAC and ion exchange resins.

This study investigated using amine modified ultrafiltration (UF) membranes to capture a PFAS mimic. Dodecylbenzenesulfonic acid (DBSA) was used instead of PFAS due to its of concentration detection using UV-Vis spectroscopy. Polyethersulfone (PES) membranes were modified using UV photopolymerization with vinyl amine monomers to create membrane adsorbers. We investigated the use of diallyldimethylammonium chloride (DADMAC), 1-vinylimidazole (1-VM), 2-[methacryloyloxy)ethyl]trimethylammonium chloride (MOETMA), and butyl acrylate (BA) as monomers. These amine polymers interact with the acidic head group of the DBSA molecules leading to adsorption. The membranes were characterized with ATR-FTIR. Static adsorption tests with DBSA were used to determine the adsorption capacity and adsorption kinetics of each membrane. A base rinse was tested to regenerate the adsorption capacity. The initial pH of the feed solution was tested to investigate the effect of solution pH on adsorption capacity. The results of this study indicate that 1-VM is a promising chemistry for PFAS adsorption.

637562 Sulfonated Poly(styrene-isobutylene-styrene) Membranes with Counter-Ion Substitution for the Separation of Water from Urea

Abstract:

The development of a low energy alternative for the separation of hydrophilic components has been studied for the purpose of recycling wastewater, such as urine, in situations in which resources like energy and clean water are limited. The other main component in urine besides water is urea, which forms strong hydrogen bonds with water making it a very challenging separation. In this study, the separation process of urea and water was completed using a polymer nanocomposite membrane (PNM) made from sulfonated poly(styrene-isobutylene-styrene) (SIBS) (81.52% sulfonation) with and without Mg^{+2}, Ca^{+2} and Ba^{+2} counter-ions. The counter-ions cross-linked into the ionic domains of the PNM, while the membrane in its acid form was used as control. Characterization techniques were applied to corroborate the presence of Mg^{+2}, Ca^{+2} and Ba^{+2}, and to identify the changes in their properties in comparison to the control. Thermogravimetric Analyses (TGA) demonstrated a major thermal stability and less water absorption in our samples, it also reduced the weight loss up to 15% and produced more temperature resistant membranes (up to 150°C higher). Atomic Force Microscopy (AFM) was performed, and morphological changes in the arrangement of hydrophilic and hydrophobic domains were observed resembling channels that could facilitate the transport of water. The permeability of urea through these membranes was studied as well. For this, the identification and quantification of urea within large water quantities was accomplished utilizing Fourier-Transform Infrared Spectroscopy (FTIR) in a range of 0.5 to 3.0% concentration. A calibration curve was developed between the absorbance and concentration of urea, which allowed us to calculate permeabilities using a diffusion cell. The permeability values were $1x10^{-6}$ for Mg^{+2}, $2x10^{-5}$ for Ca^{+2} and $4x10^{-7}$ for Ba^{+2} cm^2/s suggesting that these PNMs can be excellent candidates for the separation of urea in water. Further studies will focus on the understanding of the transport mechanism of urea, as well as the study of the permeability of water.

637646 Characterizing N,N-Dimethylcyclohexylamine (DMCHA) across Temperature Using Kamlet-Taft α, β, and π* Parameters

Abstract:

Switchable Hydrophilicity Solvents (SHSs) are potential green solvent alternatives for the extraction of high-value products. The use of SHS may reduce process energy costs associated with solvent separations and the number of required extraction steps.1 One promising SHS, N,N-dimethylcyclohexylamine (DMCHA) has been shown to radically change its hydrophilicity with the use of CO2 as a trigger in the presence of water.2 In addition to this drastic change from hydrophobic to hydrophilic, tuning of DMCHA's hydrophobicity can be achieved by small changes in temperature and can further enhance the applicability of this solvent.3 The objective of this study is to determine the Kamlet-Taft solvent solubility parameters (, , *) for DMCHA across a range of temperatures with various dissolved water concentrations to more fully understand the range of DMCHA's solvation capacity. The solvatochromic effect was measured via four commonly used solvatochromic probes (4-Nitroanisole, 4-Nitroaniline, N,N-dimethyl-4-nitroaniline, and Reichardt's dye) in a UV-Vis Spectrophotometer. The , , and * parameters were determined for 1, 3, and 5wt% water in DMCHA solutions along with Switched DMCHA (SDMCHA) and three reference solvents (water, DMSO, and Cyclohexane) at temperatures ranging from 25°C to 60°C. It was found that the polarizability (*) of DMCHA a strong function of water content and temperature - increases in water content and decreases in temperature lead to lower values of *. Furthermore, the relationship between temperature and water content is complex where larger water contents lead to a larger decrease in * across temperature. These findings illustrate the potential versatility and tunability of DMCHA as an alternative to conventional solvent based extraction methods.

637658 Green Solvent Extraction of Betulin from the Birch Bark

Abstract:

The majority of plastic products seen on the market are sourced from fossil-fuel derived polymers. The use of alternative feedstocks will allow for decreased dependence on fossil fuels as well as potentially offer other benefits such as increased polymer functionality and biodegradability. One promising bio-based monomer is betulin, a molecule commonly found in birch bark. Birch bark removed from the forest industry is potentially a plentiful, underutilized and regenerable source for bio-based plastic. This research seeks to optimize betulin extraction from dried, milled birch bark using green solvents and processes. The betulin yield and selectivity for reflux boiling (RB) with select green organic solvents and supercritical carbon dioxide assisted extractions (SFE) with green co-solvents is compared. Based on liquid chromatography-mass spectrometry analysis, the average yield of SFE was approximately 13% higher than that of RB. Precipitation of the solvent extract with water resulted in high purity solids betulin recovery. Furthermore, the betulin content in different species of birch tree is also compared. The reflux boiling products obtained from experiments under the same extraction conditions using different barks showed that the betulin concentration in locally sourced silver birch (*Betula Pendula*) is approximately 25% higher than that of locally sourced river birch (*Betula Nigra*), and *Betula Nigra* has a higher content of betulinic acid and lupeol. These results provided a preliminary understanding on the potential of green solvents in the betulin extraction and polymerization process, which will be the development basis for more rigorous analysis and larger scale experiments.

637660 Comparison of Filtration Properties and Air Flow Characteristics of Membranes and Fibrous Depth Filters for Their Application in Air Purification

Abstract:

Providing particulate-free air is a major challenge not only in the chemical industry but garnered attention during the worldwide Covid-19 pandemic last year. Filter materials are frequently used for the separation of particles from air. The choice and assembly of the filter material is critical and must be done carefully to ensure high filtration efficiencies and low pressure drops. In this work, the air flow and filtration characteristics of commonly applied depth filter materials are examined and compared with those of a membrane.

The airflow characteristics through a fibrous coarse polyester (porosity 98%, fiber diameter 25μm) and fiberglass depth filters (porosity 94%, fiber diameter 4 μm) was examined. These different filters are commonly used in ventilation systems. Filtration efficiencies were examined using spherical polystyrene latex nanoparticles (0.5 and 1 μm) as well as with potassium chloride particles. Experiments were conducted to measure the filtration efficiency and pressure drop across the different filter media at a constant face velocity of 0.07 m/s. This data was then compared to model data using single fiber efficiency models. The aerosol experiments were then expanded to include Solecta PVDF 400 membranes (porosity 46%, pore diameter 45 nm), which are commonly used in microfiltration for the removal of bacteria, fungi, fats, and microorganisms. Pressure drops and filtration efficiencies of the fibrous filters were examined and expected results regarding these parameters were found.

The glass fiber filter gives higher filtration efficiencies than the coarse material but having a higher-pressure loss. An underestimation of filtration efficiency between model data and experimental data especially for the polyester filter were found and attributed to, for example, electrostatic effects in the synthetic media. These effects from coulomb or dipole forces that can appear in synthetic media but are not covered yet by the model applied here.

The membrane was shown to have a much higher filtration efficiency than the fibrous depth filters providing very clean air but also having a significantly higher pressure drop. This research indicated different behavior of these materials. Due to the complementary advantages of both material types such as high filtration efficiency of the membrane and low pressure drop of depth materials a combination of these materials could be a promising approach for further analysis.

638180 Addressing Pfas Contamination in Blood Bank Supplies with Hydrogel Nanocomposite Sorbents

Abstract:

Environmental pollutants continue to be a threat to global human health. Persistent contaminants, such as perfluoroalkyl substances (PFAS), have been linked to a multitude of adverse health effects such as cancerous tumors, increased blood cholesterol levels and liver damage. The dominant source of exposure to PFAS is through contaminated drinking water, and accumulation has been found to occur significantly in human blood serum. Thus, high-risk groups who are receiving frequent blood transfusions are exposed to these harmful chemicals in a dual fashion, which could prove detrimental. Traditional sorbents that display an affinity for PFAS include powdered activated carbon and clay. Recently, a protein found in plasma, albumin, has been identified as the major carrier protein for PFAS in human blood. The two most widely detected PFAS in human serum are perfluorooctanesulfonic acid, PFOS, and perfluorooctanoic acid, PFOA. Thus, this work aimed to develop hydrogel nanocomposites that have the capability to remove PFOA and PFOS from human blood serum. Crosslinked acrylamide polymers were synthesized with varied crosslinking densities of 0.1 mol%, 1 mol%, and 10 mol% to evaluate potential exclusion of serum proteins. In order to incorporate physiochemical properties of sorbents known to bind PFOA and PFOS, varied amounts of dried sorbent particulate were integrated into the synthesized hydrogels. Powdered activated carbon, sodium montmorillonite clay, and bovine serum albumin were studied at loadings of 1 wt% and 5 wt% respective to total reactant weight. The synthesized hydrogels were characterized via FTIR and TGA analysis. Competitive binding to evaluate PFOA and PFOS affinity was completed in a binding matrix of pH 7.4, similar to blood serum.

638670 Development of a Continuous Liquid-Liquid Extraction of Salicylic Acid with Integrated Membrane Separation

Abstract:

The transition from traditional batch processing to flow has been a key topic in the pharmaceutical and specialty chemicals fields. In many cases, flow operation has advantages over batch in terms of control, safety, and cost. However, the implementation of flow presents different challenges including workforce training, motivating the development of a hands-on training course on the fundamentals of heat transfer, mixing, separation, microreactor assembly in flow chemistry. The work presented focuses on building an efficient flow system for the extraction of salicylic acid. Salicylic acid is an important chemical in the pharmaceutical and cosmetics industries due to its anti-inflammatory and antibacterial properties. It is also the main chemical used in the production of acetylsalicylic acid (ASA); a common over-the-counter medication known as Aspirin. The consumption of free salicylic acid is toxic to humans. Therefore, it is important to ensure that there is no residual salicylic acid in drugs such as Aspirin. In this work, batch extractions have been conducted in different solvent systems to compare their performance. Then the best-performing solvent system was used in a flow system equipped with a static mixer to enhance extraction and a liquid/liquid membrane separator. The concentration of salicylic acid was measured by taking advantage of the reaction between salicylic acid and Fe3+ to produce a violet complex ion that can be detected at 512 nm using spectrophotometry. Compared to batch, flow extractions took less time and were more efficient. In flow extractions, the addition of the static mixer enhanced mixing and showed better extraction. The static mixer enhances interfacial mixing by maximizing contact between the oil and water phases.

638464 Rare Earth Element Recovery from Coal Fly Ash

Abstract:

Rare earth elements (REE) are metallic elements that consist of fifteen of the lanthanides, scandium, and yttrium. Despite their name, REE are not rare. REE are found in common minerals, albeit at low concentrations. REE have large radii and high charges, meaning that they cannot fit into most mineral structures and there are few geological environments that support them. REEs have a wide range of applications such as magnets, wind turbines, optical materials, and electronics. Depending on the literature source, China controls 90-100% of the REE on the market. Like many metals, improper disposal of REE can be harmful to the environment through leaching into soil and water.

In this research, coal fly ash is explored as a source of REE within the U.S. Fly ash is abundant in many locations around the U.S. and has been shown to contain trace amounts of REE. Extracting REE from fly ash would address issues of supply control and price, as well as reduce leaching of those metals into the environment. Researchers at Sandia National Laboratories have developed a process to extract REE from coal fly ash using water, supercritical CO_2, and small organic acid chelators. The goal of this project is to determine if and how this process can be economically feasible at the large-scale.

Experiments will be conducted with increasing levels of fidelity to the Sandia process. Coal fly ash will be extracted with water and water plus chelator at different temperatures and atmospheric pressure. Then, experiments will be conducted at elevated pressures with nitrogen or carbon dioxide in a batch or continuous supercritical fluid reactor. These experiments will give insight to the effects of temperature, kinetics, thermodynamics, and residence time on chelators.

638492 Advanced Manufacturing of Novel Polymer Membranes Using Direct Ink-Writing

Abstract:

Abstract: Advanced manufacturing of novel polymer membranes using direct ink-writing

Nhan Khuu, Brian Leonard, Harrison Loh, Konstantinos Sierros and Oishi Sanyal

A novel membrane fabrication strategy was examined utilizing a 3-D printing strategy known as a direct-ink writing (DIW), where a viscous polymer dope solution was extruded and cast on a glass substrate. The purpose of this work is to develop alternate strategies to conventional solvent casting/doctor blading methods for making membranes. The DIW method of fabrication has two distinct advantages over the traditional techniques. – (i) DIW allows precise control of the membrane morphological properties (pore size distribution, porosity) – (ii) reduction in the amount of material (solvents/co-solvents and polymer) needed. This presentation will cover the effects of various DIW parameters on polysulfone (PSF) membrane separation properties and compare the properties of these membranes with commercially available PSF membranes. These membranes were primarily characterized based on their pure water fluxes and molecular weight cur-offs. Advanced characterization techniques like scanning electron microscopy (SEM) were employed to understand the membrane internal structure and morphology in detail. The stability of the membranes prepared were evaluated and a "low-surface tension solvent-exchange" technique predominantly used for hollow fiber fabrication, was utilized in this work to promote membrane stability. While the current work focusses primarily PSF membranes, these findings could be easily extended to other polymer membranes as well.

638498 Development of 3D-Printed Membranes for the Production of Low-Carbon Intensity Biofuels

Abstract:

Rising energy demands and the excessive consumption of fossil fuels have led to increased greenhouse gas emissions, scarcity of resources, and fluctuating fuel prices, prompting the exploration of sustainable and viable alternative energy sources. Of them, renewable bioethanol has been widely considered due to the availability of biomass and recent efficiency gains in corn-to-ethanol conversion. However, despite these gains, the efficiency of bioethanol processing remains limited, largely due to the energy-intensive distillation and molecular sieving processes currently used to purify dilute aqueous solution from fermentation. Pervaporation, a method of membrane separation, is an alternative that has the potential to address the limitations of both distillation and molecular sieving given they can demonstrate the stability, selectivity, and low capital cost required for commercial viability. Minimizing capital cost, in particular, poses a challenge to conventionally casted pervaporation membranes as their uncontrolled and relatively high thickness limits their throughput and requires larger membrane areas. In this study, a novel 3D-printing process known as electrospray was implemented to fabricate one of the first thin-film composite pervaporation membranes in order to obtain greater control over membrane thickness and improve flux without any decrease in selectivity for ethanol dewatering. Electrospray works by depositing layers of nanoscale monomer droplets in the presence of a strong electric field, which then polymerize to create a thin film. Using this method, 1, 3, and 5% polyvinyl alcohol (PVA), the benchmark standard material used for ethanol dewatering membranes, and a 10% glutaraldehyde crosslinking solution were sequentially deposited on a polyacrylonitrile (PAN) substrate. Using profilometry, crosslinked 5% PVA membranes produced through additive manufacturing were found to range from 0.404 to 1.013 microns thick for 8 to 15 layers, noticeably thinner than the 30-plus micron range typically observed for casted PVA membranes. Membrane testing using a pervaporation system is currently in progress to determine the performance of printed membranes in separating a mixture of 90% ethanol and 10% water relative to commercially available as well as conventionally casted membranes.

638575 Advancement of Pharmaceutical Manufacturing through Selective Polymorph Control during Continuous Heterogenous Crystallization: A Case Study Using Indomethacin

Abstract:

For pharmaceutical manufacturing, transferring processes from batch to continuous has the potential to more rapidly provide patients with inexpensive and higher quality medicine. In this work, under NSF Award #1555647, Indomethacin (IMC), an anti-inflammatory small molecule drug, was used as a model active pharmaceutical ingredient to achieve continuous heterogeneous crystallization onto polymer beads. IMC has multiple polymorphs, and it was the goal of this work to selectively crystalize the stable γ form and the metastable α form onto Polyvinyl alcohol (PVA, MW 89000-98000). Commercially available IMC in the pure γ form was used as the starting material. Before continuous crystallization was achieved, batch experiments were performed to determine ideal operating conditions. The crystals from each experiment were filtered and dried under vacuum at room temperature for one hour and analyzed using XRD, DSC and optical microscope. The data was then compared against literature to determine the polymorph forms that were present in the sample. Initially, Ethanol was used as the solvent and work was focused on cooling the hot solution and varying the supersaturation value, the ratio of the initial concentration of IMC in Ethanol at the starting temperature to the solubility of IMC in Ethanol at the final temperature. However, it was found that a mixture of both γ and α forms would crystalize initially and pure γ was achieved after over 5 hours as the α form slowly stabilized into the γ form. As a solvent replacement, Acetonitrile was used which, after exploring multiple supersaturation values, allowed for pure γ form to be achieved after only 30 minutes, providing significantly shorter residence time. To achieve pure α form, Ethanol was used as the solvent and it was found that introducing water as an antisolvent created a more unstable crystallization environment thus favoring the metastable α form. Both γ and α process conditions were then successfully transferred to a small-scale continuous reaction setup, illustrating a successful proof of concept for larger processes.

638577 Generating Adsorption Equations Using Symbolic Regression

Abstract:

Adsorption is relevant to a wide variety of chemical engineering problems from gas separation, pollutant treatment, and catalysis. Adsorption isotherms relate the amount adsorbed at equilibrium as a function of pressure or concentration, and are commonly expressed as equations which are either empirical or derived from theory. While several equations for adsorption have been proposed in the literature, we wonder how many more we could find. Symbolic regression is a machine learning technique that searches a space of mathematical expressions to find those that best fit a given dataset.

In this work, we use a Bayesian symbolic regression technique to analyze adsorption datasets and generate accurate and plausible adsorption isotherm expressions. Notably, we consider the effect of including prior knowledge of thermodynamic constraints to narrow the search space to find our desired model. The first thermodynamic constraint requires the loading on the surface to be 0 when the pressure is 0. The second constraint requires the adsorption isotherm to converge to Henry's Law when approaching a pressure of 0, which means that at low pressure, the limiting slope of the adsorption isotherm must be finite. We compare this technique to a genetic algorithm-based symbolic regression method that doesn't include this prior knowledge in the search. We show that we can use Bayesian symbolic regression to rediscover many common adsorption equations from experimental data, including the Langmuir and two-site Langmuir equations.

638627 Optimization of Pilot-Scale Solvent Extraction Unit for the Recovery of Scandium from "Bioacid"-Leached West KY Coarse Coal Refuse

Abstract:

Loading, scrubbing, and stripping trials using Cyanex® 272 as extractant for the liquid-liquid extraction of scandium were performed using naturally formed acid mine drainage (AMD) and pregnant leachate solution (PLS) generated from mine waste material. Results from the AMD tests showed that the recovery of Sc from AMD was highly effective using 1 vol% extractant in Orfom® diluent. The extraction efficiency of Sc was over 85% while using an aqueous-to-organic (A:O) ratio of 50:1 (v/v) and rejected 94.4% of Fe in the loading stage. Stripping using 2M sulfuric acid solution was able to effectively recover Sc from the loaded organic phase. A continuous solvent extraction system using 2 stages of loading and 2 stage of stripping was able to concentrate Sc from roughly 0.15 ppm to a 2% concentration solution.

Based on the preliminary findings from the AMD test, the process was assessed on Sc extraction from PLS obtained from mine waste material. The PLS was obtained from leaching roasted West Kentucky Coal Seam 13 coarse refuse with sulfuric media generated through oxidation of fine pyrite with *Acidithiobacillus ferrooxidans*. The acidity of the plant-generated "bioacid" was at a pH of 1.56. After leaching at 200 g/L solid concentration under 75 °C for 5 hours, the pregnant leachate solution contained 0.65 ppm of Sc with a final pH of 2.811. Majority of the Fe(III) was neutralized during the leaching process due to the elevated pH. Loading trials were performed using various A:O ratios, and a 50:1 aqueous-to-organic ratio was found to be optimal for scandium loading and limitation of iron loading. However, accumulation of Al was observed in the organic phase provided due to the high A:O ratio. Scrubbing trials were carried out to selectively remove Al ions using NaOH solution. An organic layer of iron precipitates formed during the scrubbing tests as a result of the presence of Fe in the organic. Further work is suggested to 1) generate the leachate solution at lower pH so that the pH of the PLS is below 2.0, and thus reducing the loading efficiency of Fe in the loading stage and 2) reduce ferric ions to ferrous ions to prevent Fe loading into SX circuitry. Stripping trials were not re-tested at this time, as similar initial pH and Eh constraints of the feed suggest stripping trial data from AMD work will continue to hold.

637848 Selectively Coating Specific Chiralities of Single Walled Carbon Nanotubes Using Co-Surfactant Solutions

Abstract:

Single walled carbon nanotubes (SWCNT's) generally contain a mixture of different (n,m) chiralities, and must be separated from the overall mixture to be used in different applications. Previously we have shown that specific co-surfactant ratios of sodium dodecyl sulphate (SDS) and sodium deoxycholate (DOC) can be utilized in the separation of specific chiral species. Using these different ratios of SDS to DOC it is possible to selectively coat specific chiralities with DOC among unsorted SDS coated SWCNT's. Here, we have utilized dialysis as a means to understand the time needed for the DOC coating of each chirality in the mixture. Two different dialysis setups were used; Slide-A-Lyzer MINI Dialysis Devices and Mini Dialysis Kit. The Slide-A-Lyzer contained cups to hold the solution which could then be inserted into a conical tube containing the co-surfactant buffer solution whereas the Mini Dialysis Kit contained an inverted floating tube with the dialysis membrane located on the cap of the tube. The rate of replacement of surfactants from SDS to DOC on the surface of SWCNTs proved to be different for both the dialysis methods. Furthermore, the time required to coat a specific chirality was lowered by adjustments made in the dialysis procedure such as replacing the buffer frequently and using a stir plate.

637943 3D-Printed Mixers Demonstrated for CO_2 Capturing

Abstract:

CO_2 emission is a major environmental challenge and one of the main contributors to global warming. A significant fraction of it originates from the combustion step as those used in power generation plants. In this process, amine-based absorption is the most technologically matured CO_2 capturing technology to date. Still, this process has some fundamental limitations. To substantially improve energy efficiency and reduce capital and operating expenditures, the capabilities of 3D manufacturing to fabricate non-conventional static mixers are explored to enhance the CO_2 capturing technology. Along the theme of this motivation, a multidisciplinary project was established and executed by six undergraduate chemical and mechanical engineering students. Although a major part of this project is related to education, two experimental facilities have been established, which are essential for conducting high-quality research and meeting the CO_2 capturing design requirements. The first is an air-water setup used to experimentally evaluate the flow hydrodynamic, gas-liquid bubble breakage, and measure the pressure drop. The second facility is a microreactor, which is a powerful laboratory tool for kinetics studies and screening conditions and solvents. This setup is used to understand the CO2 application requirements, and more importantly, to establish an experimental benchmark to evaluate the invented and 3D-printed mixers designs. Multiple 3D-printed mixers are fabricated and evaluated using these two setups. The speed from which a design was created, fabricated and tested, suggests that many design iterations can be lead to validated novel mixer designs with low resource and budget requirements. Progress reached so far will be presented along with learning from using this experimental-driven design approach enabled by 3D manufacturing.

Acknowledgment: This work was made possible by funding from UREP27-097-2-028 (Undergraduate Research Experience Program) from Qatar National Research Fund (QNRF).

638484 Investigating Vanadium Supported Zeolite for Direct Air Capture of CO2

Abstract:

Direct Air Capture (DAC) of CO2 is an engineering challenge that has been studied for about two decades. The process is primarily motivated as a possible rectification to the anthropogenic increase in atmospheric CO2. Catalytically converting the captured CO2 to value-added products further incentivises CO2 capture for a more sustainable economic trajectory. Finding an ideal DAC sorbent involves navigating through thermodynamic and kinetic impediments such as minimizing energy input for regeneration and maximizing the rate of CO2 adsorption given the low (400 ppm) concentration of CO2 in air. Materials such as CaO, amines, hydroxides, and metal organic frameworks have been studied, but there is still room for improving working capacity and energy efficiency, in particular during the CO2 desorption step.

To address these challenges, it is hypothesized that vanadium dispersed in a zeolite may be a material that exhibits desirable properties for DAC of CO2. Vanadium offers facile access to multiple oxidation states, has a nucleophilic surface, and can generate oxygen atom defects, all of which assist in binding and activating CO2. However, vanadium atoms in solid state materials have been shown to migrate, agglomerate, and volatilize. Stabilizing vanadium in a zeolite matrix could prevent these problems, but vanadium is known to catalyze dealumination which destroys the alumino-silicalite zeolite framework. Much of the sorbent would also remain inaccessible to the CO2 because of diffusion limitations within the micropores. To this end, it is proposed that dispersing vanadium in a boron-silicalite zeolite with meso/macro porosity is a potential candidate for DAC. By replacing aluminum with boron, it is expected that the same advantages of stabilizing the metal on a zeolite matrix are achieved, and the zeolite framework is also protected from potential degradation by vanadium. By introducing meso/macro porosity, the material can be tuned for CO2 adsorption, thereby decreasing the barrier to diffusion. Additionally, CH4 oxidation can be catalyzed on or nearby the active site where CO2 is adsorbed. A controlled CH4 oxidation would release heat into the system where CO2 is bound, effectively making the system able to convert a potent greenhouse gas CH4 into a lesser potent CO2 and meanwhile reducing the energy input for catalyst regeneration. Thereby, the proposed structure presents itself as a promising sorbent for direct air capture of CO2.

AUTHOR INDEX